Behind the Lines of Stone

The Social Impact of a Soil and Water
Conservation Project in the Sahel

Nicholas Atampugre

Oxfam
UK and Ireland

© Oxfam (UK and Ireland) 1993

A catalogue record for this book is available from the British Library

Front-cover photograph by Jeremy Hartley

ISBN 0 85598 257 8 (hardback)
ISBN 0 85598 258 6 (paperback)

Published by Oxfam (UK and Ireland)
274 Banbury Road, Oxford OX2 7DZ
(Oxfam is registered as a charity No. 202918)

Designed and typeset by Oxfam Design Department NY285/PK/93
Printed by Oxfam Print Unit

Typeset in Melior 10/12pt
Cover PMS 172 & 329
Printed on environment-friendly paper

Contents

List of tables   v
List of figures   vi
Contributors   viii
Acknowledgements   ix
Abbreviations and a note on currency   x

Introduction   xi
    Burkina Faso   xi
    Projet Agro-Forestier   xiv
    The 1992 evaluation   xv
    The book   xix

1 Yatenga: a battle for survival   1
    Population and environment   2
    The farming system   9
    Water resources   11
    Seasonality   12
    Socio-economic infrastructure   15
    Migration from Yatenga   22
    Comment: the challenge facing NGOs   24

2 External agency intervention in Yatenga   25
    Learning from the past   26
    Recent NGO involvement   27
    PAF: the early years   34
    Comment: some lessons from the early years   46

3 The scope of PAF today   53
    Objectives and priorities   53
    Areas of activity   57
    Geographical zones of intervention   64
    Comment: competition or co-operation?   66

4 Laying the foundations of development   72
    The impact of *diguettes*   72
    PAF's role in *diguette* construction   84
    Beyond *diguettes*: PAF's complementary activities   91
    Comment: how strong are the foundations?   101

5 Coping with hunger   109
    Survival strategies   109
    The socio-economic impact of PAF   114
    The limits of PAF's role   117
    PAF's impact on women   121
    Comment: a targeted, sustainable enterprise?   132

6 Past, present, and future   136
    PAF and participatory methodology   137
    Institutional learning   139
    Technological development   141
    Institutional structures for participation   143
    PAF's claim to cost-effectiveness   149
    Reaching the most disadvantaged   151
    Looking into the future: PAF and the State   154
    Conclusion   160

References and suggestions for further reading   163

Index   165

List of tables

Table 1.1 Health infrastructure in Yatenga Province 16

Table 1.2 Distribution of households according to type of medical care 18

Table 1.3 Distribution of households according to source of water 19

Table 1.4 School-attendance rates (7-to-14 year age group) 20

Table 1.5 Educational levels of women in Yatenga 21

Table 2.1 Externally-funded organisations with offices in Yatenga 28

Table 2.2 Impact of diguettes on cereal yields 48

Table 3.1 Agencies with which PAF co-operates 56-7

Table 3.2 PAF's zones of intervention, 1992 65

Table 4.1 Numbers of households growing particular crops on treated plots 75

Table 4.2 Number of households involved in diguette construction 78

Table 4.3 Households according to type of agricultural equipment owned 80

Table 4.4	Households' ranking of problems confronted in diguette construction	81
Table 4.5	Number of households with knowledge of diguette-construction techniques	84
Table 4.6	Households according to type of diguette constructed	88
Table 4.7	Households according to length of knowledge of PAF	89
Table 4.8	Households according to source of knowledge about PAF	89
Table 4.9	Households according to source of knowledge about soil-conservation techniques	90
Table 4.10	Households according to prioritisation of soil conservation and plant regeneration	91
Table 4.11	Number of farmers trained by PAF, and nature of training	96
Table 4.12	Distribution of households according to manner of diguette construction	97
Table 4.13	Households according to use of revolving grain stock	98
Table 4.14	Households according to capacity to repay grain loans	99
Table 5.1	Women's experience of the 1990/91 hunger	110
Table 5.2	Households' coping mechanisms during 1990/91 hunger	111
Table 5.3	Distribution of households according to sources of income	113
Table 5.4	Cereal balance (tonnes), 1991/92 farming season	116
Table 5.5	Evolution of market gardening in Yatenga province: production (tonnes), 1981-1991	118

Table 5.6	Distribution of women according to benefits from or inconvenience of involvement in PAF activities 122
Table 5.7	Distribution of women according to type of PAF activity with which they are involved 127
Table 5.8	Distribution of women according to income-generating activities 128
Table 5.9	Nature and source of women's access to farms 129
Table 5.1	Distribution of women according to use of resources 129
Table 6.1	PAF's budget - 1982 to 1992 150

List of figures

Fig. 1	North and West Africa, showing Burkina Faso and the extent of the Sahel xii
Fig. 2	Burkina Faso: provinces and administrative centres xxiv
Fig. 3	Yatenga Province: PAF's zone of intervention 3
Fig. 4	Yatenga Province: relief 6
Fig. 5	Yatenga Province: vegetation 7
Fig. 6:	Yatenga: average annual rainfall, 1921-1989 11
Fig. 7	The water tube 40

Contributors

This book is based on the findings of the following people, who acted as researchers in the field, and consultants in the early stages of the study.

Sylvester Aweh (teacher of English): First synthesis of drafts in English

Albertine Darga (geographer): The environmental problems of Yatenga

Rasmané Kaboré (agronomist): PAF's role in soil and water conservation

Eloi Ouédraogo (statistician — National Institute of Statistics and Demography): Director and co-ordinator of field enquiry

Seydou Ouédraogo (journalist): NGOs in the province of Yatenga

Maliki Sawadogo (agro-economist): Socio-éeconomic impact of PAF

Pauline Yabré (teacher): Socio-economic impact of PAF (focus on women)

Acknowledgements

Many people were involved in the various stages of this book's development. Its basic outline benefited from the comments of Nigel Twose of ActionAid, John Rowley and Michael Butcher of Oxfam (Oxford), Alice Iddi-Gubbels and the late Mamadou Koné of Oxfam (Ouagadougou), and Mathieu Ouédraogo, Sanaa Issaka, and Daouda Ouédraogo of PAF, based in Ouahigouya. The process of selecting contributors to conduct field research would not have been possible without the active support of Oxfam's team in Ouagadougou in general, and Alice Iddi-Gubbels in particular. The draft was improved by the comments of the Oxfam staff in Ouagadougou and Oxford, Peter Wright, the first co-ordinator of PAF (currently working in Koudougou, Burkina Faso), and Nigel Twose of ActionAid, Chris Roche of ACORD, Charles Abugre, formerly of ACORD, and George Kararach of the University of Leeds. It was their critical comments which finally brought this book to its current shape. However, the main responsibility for the production of the book fell on me in my capacity as an independent consultant and director of the PAF book project. Any lapses in this book are my responsibility.

Nicholas Atampugre
London
September 1993

Abbreviations used in this book

AFVP: *Association Française des Volontaires du Progrès* (Association of French Volunteers for Progress)
BSONG: *Bureau de Suivi des Organisations Non-gouvernementales* (National Secretariat for the Co-ordination of Non-Governmental Organisations)
CNR: *Conseil National de la Révolution* (National Council of the Revolution)
CRPA: *Centre Régional de Promotion Agro-Pastorale* (Regional Centre for the Promotion of Agro-Pastoralism)
CIEPAC: *Centre International pour l'Education Permanente et l'Amenagement Concerte*
FCFA: *franc de la Communauté Financière Africaine*
FDR: *Fond du Développment Rural* (Rural Development Fund)
GERES: *Groupement Européen de Restauration des Sols* (European Society for the Restoration of Soil)
GNP: gross national product
NGO: non-governmental organisation
ORD: *Organisme Régional de Développement* (Regional Development Organisation)
PAE: *Projet Agro-Ecologie* (Agro-Ecology Project)
PAF: *Projet Agro-Forestier* (Agro-Forestry Project)
PNGT: *Programme National de Gestion des Terroirs* (National Programme of Land Management)
PVNY: *Projet Vivrier Nord Yatenga* (North Yatenga Food Growing Project)
SDID: *Société de Developpement International Desjardins* (Desjardins International Development Agency)
Six S: *Se Servir de la Saison Sèche en Savane et au Sahel* ('Making Use of the Dry Season in the Savanna and the Sahel')

A note on currency

The currency of Burkina Faso is the FCFA (franc de la Communauté Financière Africaine), which since 1948 has been pegged to the French franc at the rate of 50 FCFA: 1 French franc. In 1992 the average exchange rate against the US dollar was 265:1.

Introduction

Oxfam (UK and Ireland) is primarily a funding agency providing financial assisance to intermediary implementing agencies which include state ministries and other voluntary agencies (both national and international) working on small development projects. In the West Africa sub-region, however, Oxfam does have four operational projects where it is directly responsible for planning and implementation. One of these projects is located in Yatenga, in the north of Burkina Faso. It is the *Projet Agro-Forestier*, or Agro-Forestry Project, a soil and water conservation project more commonly known by its acronym: PAF.

Burkina Faso

Burkina Faso was a French colony until 1960, and was known as Upper Volta until it changed its name in 1984. It is located on a plateau that rises gently from an elevation of 240 metres (800 feet) in the south-west to 350 metres (1,150 feet) in the north-east. It is a small, land-locked country in the heart of West Africa, covering approximately 274,200 square kilometres (105,800 square miles). It shares borders with Côte d'Ivoire (in the south-west), Mali and Niger (north-west and north-east), Togo and Benin (south-east), and Ghana (south).

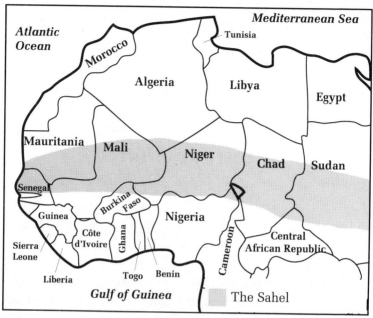

Fig. 1: North and West Africa, showing the position of Burkina Faso and the extent of the Sahel

Its population stands at about 9.5 million inhabitants (1992 estimate) and was growing at approximately 2.6 per cent p.a. between 1980 and 1985. The Mossi plateau has the highest population density and accounts for about half of the population. The average population density is about 32.5 inhabitants per square kilometre, but densities in the Mossi plateau are more than twice the national average. The urban population is estimated to be growing at 5.4 per cent p.a. since 1980. There are about 60 ethnic groups, many of them small minorities. The Mossi, who mostly live in the centre of the country, constitute more than 50 per cent of the total population. The rest include the nomadic Fulani in the north, the Dioula in the west, the Gourounssi, the Dagara, and the Gourmantché.

Burkina Faso derives most of its external revenue from exports of cotton and livestock (the latter providing about one-third of the country's export earnings), remittances from

Introduction

migrants in Côte d'Ivoire (estimated to number about two million people), and foreign aid. In recent years contributions by non-governmental organisations (NGOs) to the national economy have become an important source of foreign earnings, even if they are not channelled to the national treasury. In 1986, the contributions of NGOs amounted to over US$20 million: 8.5 per cent of total external assistance. In 1987, this rose to US$33 million: 15 per cent of total external assistance. NGOs' total contribution for the period 1986-1990 has been estimated at US$127 million (at 1990 rates of exchange). In contrast, Burkina Faso's external debt has been estimated in the country's 1991-1995 development plan at US$722 million (again, at 1990 rates), of which 63.5 per cent is owed to multilateral creditors, 35.8 per cent to bilateral creditors, and 0.7 per cent as supply credit.

Agriculture (farming and livestock rearing) is the backbone of the national economy. It provides employment and revenue for 90 per cent of the population; 6.3 per cent of the total population are engaged in livestock rearing. The livestock sector contributed 8.5 per cent to GNP, according to the 1985-1990 development plan, a 20 per cent decline compared with the 1970s. The contribution of the primary sector to the GNP was estimated at 31.5 per cent for the period from 1985 to 1990.

Though slight increases in cereal production have been recorded in certain years (5.4 per cent in 1985/86 and 3.4 per cent in 1988), persistent annual food deficits characterise Burkina Faso's agricultural system. Since the famine years of the 1970s, food production has remained insufficient to cover national food requirements. From 1977 to 1981, food-production deficits ranged from 150,000 tonnes in 1977 to 93,700 tonnes in 1980. Between 1984 and 1989, all the provinces of the Sahel zone experienced cereal-production deficits, ranging from 72,294 tonnes in 1984 to 52,535 tonnes in 1989. During the 1989-1990 agricultural year, the annual food deficit stood at 52,000 tonnes.

The secondary sector covers a wide range of activities, including housing construction, public works, small-scale industries, the self-employed sector (arts and crafts), transportation, trade, and tourism. The mining and manufacturing sector has recorded a slight increase in recent years: it represented 18 per cent of the GNP in 1985 and 23 per cent in 1987. Gold deposits have been discovered in the north of the country; the mines are of the open-cast type and have become a source of income for private gold diggers.

The country's road network links it to its neighbours; the best route is the road link between Burkina Faso and its major trading partner in the region, Côte d'Ivoire. With the exception of roads which connect provincial capitals to the national capital, most of the country's road network remains untarred. There is also a railway line which used to be co-managed by Côte d'Ivoire and Burkina Faso, until 1988, when a National Railway Authority was created, called the Société de Chemin de Fer du Burkina (SCFB). The line principally carries heavy freight and livestock from Ouagadougou through Bobo-Dioulasso, the nation's economic capital, to the sea port of Abidjan in Côte d'Ivoire.

It is within this national context that Oxfam's intervention in Burkina Faso and the work of PAF, its soil and water conservation project, operate.

Projet Agro-Forestier (PAF)

In 1992, PAF was 13 years old. It had begun in 1979 simply as a research project, and had developed by 1982 into a large-scale operation actively involved in helping farmers to conserve, protect, and develop their natural-resource potential. Although 13 years is a short period in rural development, the project has succeeded in making its presence felt, both locally and internationally. At the local level, its simple instrument for determining contour lines, the water tube, has transformed the struggle to harness run-off water for productive use within an area of declining rainfall. Nearly all agencies involved in soil and water conservation have found the use of the water tube in the construction of stone *diguettes* or *bunds* to be an effective strategy for halting the process of soil erosion.

Introduction

Internationally, PAF has been the subject of discussion at meetings which review strategies for promoting productive activity in fragile eco-systems. Working papers, seminar documents, and newspapers have all cited the technique (even if they have not always attributed it to PAF) as a worthy example of innovation. Visitors from within the country and from beyond, travelling to Yatenga to see the technique, have been encouraged to adapt it to their local conditions. Since 1989 the project has been filmed, photographed, and repeatedly written about, to such an extent that Oxfam's programme in West Africa became more or less synonymous among Oxfam's supporters with the 'Magic Stones' project. Stone *diguettes* presumably earned the accolade 'magic' because of their effectiveness in slowing down or halting erosion. In fact, in the five years since 1988, PAF has been perceived to be one of the most successful soil and water conservation projects in the region. However, the people behind the stones have almost always been forgotten, and how exactly the *diguettes* have improved the material well-being of local farmers has so far not been well established.

The 1992 evaluation

Oxfam recognises the progress made by the project over the last ten years. But it is also conscious of the fact that it is the beneficiaries of development who can best say how they have benefited: they are the ones who know whether the shoe fits or hurts, and where. Oxfam perceives the process of development as being, in part, a learning exercise. In carrying out the field study which led to the production of this book, the guiding approach was to learn from the experience of the project and share it with those for whom it may be relevant. Oxfam had never systematically documented and published in detail, and for a wider audience, the experiences and lessons of PAF. A decade, though short, constituted an appropriate time frame to begin to look back and assess how far the project had progressed. Such an assessment was also deemed crucial for planning the project's future.

In deciding to document the experiences of PAF, Oxfam's objective was not to publicise the project, which had already been sufficiently reported by outside researchers. Indeed, it could be said to have been over-publicised, with the consequent danger of creating illusions which might be destroyed if the evaluation discovered signs of serious inadequacy. The decision to document the experiences of PAF was guided by a desire to ask questions which external researchers might hesitate to ask for fear of offending the agency. In commissioning an assessment of the project, Oxfam was prepared to consider the possibility that perhaps the success of PAF had been exaggerated. Perhaps the project had not really fostered any critical awareness among farmers of how best to harness their immediate natural resources to improve their living conditions. If development is about improving the life of people, then who have benefited — and how? Is there anything to show that the project has contributed in any way to make life better for some of the impoverished people of Yatenga? Could it not be the case that, after all, the poor who were intended to be the beneficiaries have turned out to be the losers?

The research team

Oxfam is aware that questions like these are not easy to answer, especially in an environment where generalised impoverishment is the norm, and in a project whose initial objective was to promote the growing of trees. Yet it believes that even if these questions are not exactly the most appropriate, they are nevertheless worth exploring. In deciding to carry out the study, it recognised the difficulties inherent in asking such critical questions using resources internal to the project and to Oxfam. In the words of a Mossi proverb, 'It's not easy to detect the smell of your own mouth'. In recognition of this fact, the agency commissioned independent observers to help it to ask awkward questions and raise issues for debate. A multi-disciplinary team, consisting of researchers from Burkina Faso, able to communicate directly in Mooré, the local language of the

Introduction

people of Yatenga, was assembled for the purpose. Team members were selected on the basis of their proven skills, preparedness to engage in a dialogue with members of the village community in the project area, and their limited knowledge of or association with the project or Oxfam. The choice of team members and the design of the study (including a determination of the questions to be asked) was undertaken by the present writer, a northern Ghanaian researcher with experience of conducting similar exercises, and a practical appreciation of life under semi-arid conditions. The author, who had no previous involvement with Oxfam, had the job of finally pulling together the various strands of the study into this book.

The respondents

The investigation targeted rural people within the project's area of activity. The purpose of this was to obtain comprehensive testimony from the intended beneficiaries. As a proverb in northern Ghana points out, 'It is only the person on top of the sheanut tree who can say whether its fruit is sweet or sour'. As a result, the book incorporates all the relevant testimonies of farmers and project personnel. The study recognised the limited scientific validity or representativeness of oral testimonies. This is particularly the case in Yatenga, which is so saturated with external agencies and researchers that local people seem to have a ready answer to every question. In recognition of this limitation, the study administered questionnaires not only to a representative random sample of 1,209 households in villages where the project focuses its activities ('pilot villages'), but also in villages where the project has had no involvement, or has scaled down or terminated its activities ('test villages'). The quantitative study was designed and directed by staff of the National Institute of Statistics and Demography in Burkina Faso, acting in their individual capacity. (Any inadequacies in the quantitative study are the responsibility of the individuals rather than the institute.)

The methodology

In deciding to administer written questionnaires, the limitations of such a technique, especially in non-literate rural settings, were carefully considered. The notion that quantitative data obtained from questionnaires are necessarily more valid than oral testimonies has been shown, by writers on development issues such as Robert Chambers, to be fundamentally flawed. What appears to be a simple and clear question can and does have several meanings and is subject to various interpretations by respondents. However, with all their inherent design limitations, when questionnaires are correctly administered, the results can often serve as useful pointers. Any extreme variance revealed as a result can point to areas for further investigation. It is this variance, rather than the quantitative figures, which makes them useful. The tables presented in this study, produced from the field enquiry, are therefore mainly supplementary to oral testimonies.

The study was not designed to be a project evaluation in the usual sense of the word. There was no intention to make any cost/benefit analysis or draw up a list of recommendations. Standard cost/benefit analysis tends to ignore the role of the broader social and institutional environment, and does not usually indicate the ways in which the beneficiaries have been affected and why. For example, in 1989 a World Bank team calculated the internal rate of return to PAF's investment, using conservative assumptions, to be 37 per cent. After considering the initial cost of treating a new hectare of land as sunk or unrecoverable, the return on investment to farmers was estimated to be as high as 147 per cent. The team therefore concluded that the return to PAF was at worst quite respectable and at best extraordinary. However, did such a conclusion necessarily mean that households in the project are now 147 per cent better able to meet their nutritional needs? Did that mean that the people of Yatenga no longer face a threat of famine? This is unlikely, given the experiences of households during the 1990/91 farming year, in which the

Introduction

people of Yatenga thought they were heading for a repeat of the famine of the 1970s, when Burkina Faso was severely affected by the drought and famine which swept across the Sahel.

The book
Structure and argument

This book is divided into six chapters. **Chapter 1** discusses the geography, natural resources, and socio-economic infrastructure of the province of Yatenga. The recollections of elderly people are used to highlight the problem of environmental degradation within Yatenga. The vegetation, relief, and soil types, the farming system, water resources, rainfall patterns, and the seasonal character of life in Yatenga are also described. There is an overview of the socio-economic infrastructure of the province (roads, health, water, and education), and an account of the problem of migration. The purpose of examining these issues is to remind readers of the interrelated nature of problems which confront the people of the area, as well as to set the framework within which external agencies intervene. The chapter ends by asking whether, in the light of the problems, NGOs can succeed in helping to improve the living conditions of poor communities.

Chapter 2 examines interventions by external agencies in Yatenga. It begins with a historical overview of the earliest external involvement in the province, which lays the basis for examining more recent NGO intervention in Yatenga. Five operational projects with offices in the province are briefly described. These are the Agro-Ecology Project (PAE), the Association of French Volunteers for Progress (AFVP), the Desjardins International Development Agency (SDID), the North Yatenga Food Growing Project (PVNY), and the 'Six S' Movement. Although this book is about PAF, it is important to emphasise that the project does not operate in a vacuum. The brief overview of other projects provides an opportunity to ask if PAF really is unique. Furthermore, the involvement of so many external agencies in one province raises certain issues which are discussed later.

The description of other projects is followed by a detailed account of PAF itself. The origins of the project are traced and its early years reviewed, highlighting in particular the water tube and the shift from the promotion of tree planting with *diguettes* to the use of *diguettes* for improving crop yields. Key lessons of this early history are discussed.

Chapter 3 presents a detailed account of PAF from the point in 1985 when local staff assumed full responsibility for the project. The aims, philosophy, strategy, areas, and zones of intervention are presented. The specific activities undertaken by the project — afforestation, animal confinement, compost production, training, research and development — are described, in addition to its traditional area of activity, the promotion of *diguette* construction. This detailed review is intended to lay the basis for assessing, in Chapters 4 and 5, the contribution that the project is making to development within its pilot villages. To conclude Chapter 3, there is a discussion of the problems that arise when so many NGOs are concentrated in one province, and the dilemmas that are raised by multi-agency intervention in similar areas of activity. Against this background, can PAF, like so many other projects operating within the province, claim to be making an impact on the problem of hunger, an issue central to the life of the people of the province?

Chapter 4, entitled 'Laying the foundations for development', examines PAF's contribution in its areas of intervention. It looks at *diguettes* and the range of activities that the project is involved in, and asks if they are improving rural households' chances of getting a better harvest. It examines evidence for farmers' awareness of the impact of *diguettes*, the potential that they offer for crop diversification, and their impact on the process of natural regeneration of vegetation cover. In the light of the above, the scale of *diguette* construction is considered, and the problems confronting farmers in *diguette* construction are examined, together with possible solutions.

The specific contribution of PAF to *diguette* construction is assessed and the project's other activities — soil improvement, animal confinement, training, and extension

Introduction

work — examined. The work of PAF with village groups is also examined in the chapter, focusing in particular on the contribution made by the revolving stock scheme, meant to help poor households. To conclude the chapter, the issues raised earlier are summarised in a discussion of the nature of the development foundations that the project has laid and the role that PAF has played. The chapter ends by asking whether *diguettes* by themselves are a firm foundation for development. If they are, the people of PAF's pilot villages in particular, and Yatenga in general, should find it easier to cope with lean years — the subject of the next chapter.

Chapter 5 begins with a discussion of the survival strategies of the people of the province, and the limited options available to them. The socio-economic impact of PAF is examined, highlighting in the process the difficulty of assessing the socio-economic impact of a soil and water conservation project. Many of the areas which affect people's lives directly are beyond the scope of the project's intervention. PAF's non-involvement in areas which could benefit farmers more directly is, however, identified as a weakness, and its implications for the process of migration out of the province are examined. The primary focus of this chapter, however, is the impact of the project on women, on the assumption that if a project has improved the living conditions of people, then women should be the first to be aware of it, as a result of improvements in the well-being of their children.

The chapter concludes with a summary of the main issues discussed earlier, which aims to help the reader to draw more general conclusions to answer the question: has the decade of PAF's contribution made a significant impact on the life of the people? In the process, the project's overall contribution is discussed within the context of the unchanged situation of lean years, interspersed with occasional bumper harvests. There is also a discussion of the project's role in strengthening local problem-solving capabilities through farmers' own resources. There is an attempt to find out what — apart from *diguettes* — the project has left by way of a legacy for the people of Yatenga. The view expressed in the chapter is that there is little else. The probable reasons for this are the subject for discussion in the final chapter.

Chapter 6 takes a retrospective look at PAF and attempts a projection into the future. The aim is to discuss in more general terms the claim of the project to a particular identity. Its claim to a participatory methodology is examined critically — with due recognition of the difficulty that exists in defining precisely what a project adopting such a methodology should look like. It is observed that the absence of systematic documentation has made the project's contribution difficult to define. The reasons for this situation are explored, and an attempt is made to explain the limited contribution that the project has made to local technology development.

This chapter identifies the need to develop institutional structures which might ensure a more effective involvement of local people in the running of the project. It also looks at the contribution that PAF is making towards local institution-building, as well as its claim to being a cost-effective project. Similarly, there is a review of its claim to target poorer households, bearing in mind the difficulty that exists in carrying out an identification of precise strata in rural society. This is necessary if such well-meaning aims are to be translated into clear, well-defined, and properly targeted activities. The discussion asks whether, in the light of the severity of the problem of environmental degradation, improving the overall environmental situation should not take precedence over reaching out to mainly poorer households.

The second section of the final chapter explores the future of the project. It does so by reviewing the project's programme of transition from 1992/93 to 1996/97, which has been designed to integrate the project's work into the national programme of village land management (PNGT). The history of PNGT is reviewed, and its future discussed. It is necessary to do so because the future of PNGT will to a large extent influence the shape of PAF's intervention in the province of Yatenga. The book ends with the prediction that the project's future will depend on whether it has learnt the relevant lessons of its history, especially how it can give the local

Introduction

people control over their own destiny, as well as its ability to secure funding to carry out the activities it plans to undertake.

The book's style and audience

The book is written in a journalistic but analytical style, drawing on anecdotal material and actual quotations from the testimonies of rural farmers and project officials, as well as quantitative data gathered by the field study. The objective in adopting such a style is to communicate to a wide readership the story of PAF and the struggle of the people of Yatenga to improve their lives. At one level, the book aims to inform non-specialist readers who may be more interested in the daily lives of the people of Yatenga and how they are struggling to improve their well-being with PAF's assistance. For this purpose anecdotal information is used extensively. At another level, the book hopes to interest development workers by raising the problem of how to translate familiar NGO rhetoric into practice. Competition and co-operation among NGOs, and the difficulty of ensuring that a project's activities have a more measurable and provable impact on the life of beneficiaries, are issues which remain unresolved. In general, the book prefers to raise issues rather than offer solutions. This approach is based on the view that solutions have to be worked out within a specific context, taking into account social realities. The best that can be done in many circumstances is to take stock constantly, and to document the lessons learnt so that they could be of use to others. For this reason, throughout the book, an attempt is made to present both sides of an argument.

In conclusion, it should be stated that the author and the research team were given unrestricted access to all the documentation concerning PAF, and the opportunity to visit and interview a wide range of people in the project's pilot villages and the test villages. The results of the evaluation were discussed at length and in depth with Oxfam staff members based in Ouagadougou and Oxford, but the views expressed in this book are those of the author, and are not necessarily endorsed by Oxfam (UK and Ireland).

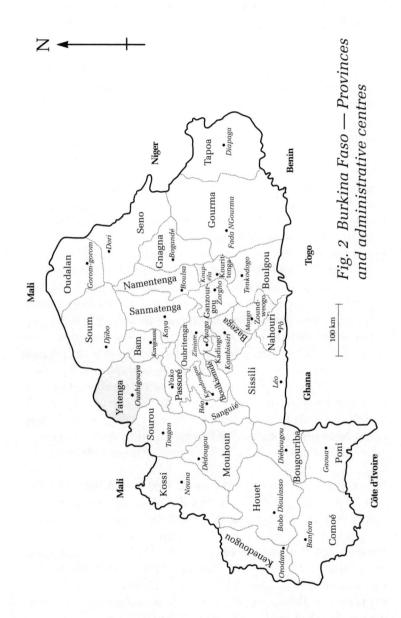

Fig. 2 Burkina Faso — Provinces and administrative centres

1

Yatenga: A battle for survival

'The seasons were good and the harvest was abundant. The environment was benevolent, and everyone knew they could savour the fruits of a few months' work in the fields. As the years went by, our luck began to turn. The rains became rare and inconsistent, the ponds dried up, and the trees died. The lions, buffaloes, and antelopes which thirty years ago used to roam these parts are nowhere to be seen. The only animals we see today are rabbits. Even birds like the nomwalga are now rare, because there are no big trees for them to perch on. Our environment has become desolate.' (Boureima Sawadogo, 92 years old)

Yatenga Province is in the extreme north of Burkina Faso, between latitudes 13° 00' and 14° 15' north, and longitudes 1° 45' and 3° 00' west. It is bounded to the north by the republic of Mali, to the south by the province of Passoré, to the east and south-east by the provinces of Soum and Bam, and to the west by the province of Sourou. The Yatenga Province has a surface area of 13,222 square kilometres, divided into 19 administrative districts. It forms part of what is known as the Mossi plateau, and is populated by various ethnic groups.

Population and environment

There are 598 villages in the Yatenga Province, with a total population of 624,000 inhabitants — about 9.4 per cent of the national population. The resident population is about 586,000 — 8.4 per cent of the national population. Yatenga is sparsely populated in the north and densely populated in the centre. The population density of the province is estimated at 48 inhabitants per sq km, although (according to the Second Five-Year Development Plan, 1991-1995) densities around Ouahigouya, the provincial capital, are as high as 100 per sq km. The people of the province are predominantly Mossi (50 per cent). The rest include Silmi-Mossi (descendants from inter-marriage between Fulani and Mossi), the Fulani, the Dogon, the Samo, the Rimaibé, and the Menancé. The active labour force of the province was estimated in 1985 at around 293,000, of whom 146,000 are engaged in the rural economy and 785 in the public sector, working for wage incomes. For the vast majority of the people of the province, obtaining their food and other basic requirements involves a bitter struggle with a hostile environment.

Concern for environmental degradation, and deforestation in particular, was expressed after the disastrous drought and famine of 1970 to 1973, which affected countries in the Sahelian region of West Africa. Although drought and famine are not new to the people of Yatenga, it was the first time the scale of the disaster in the region had attracted international attention. As Harouna Ouédraogo (M, 85 years) of Ranawa recalls, 'In 1914, there was a terrible famine which killed many people. I swear at that time, people walked to Bobo-Dioulasso, seizing food along the route and risking their lives to provide food for their dependants. Some years later, there was another famine which we named the famine of Naaba Koabga (after the chief who ruled us). After this, there was another famine in 1970, but this time the white people came to our aid.'

Although international agencies, including Oxfam (UK and Ireland), played an important role in providing relief during

Yatenga: A battle for survival

Fig 3 Yatenga Province: PAF's zone of intervention

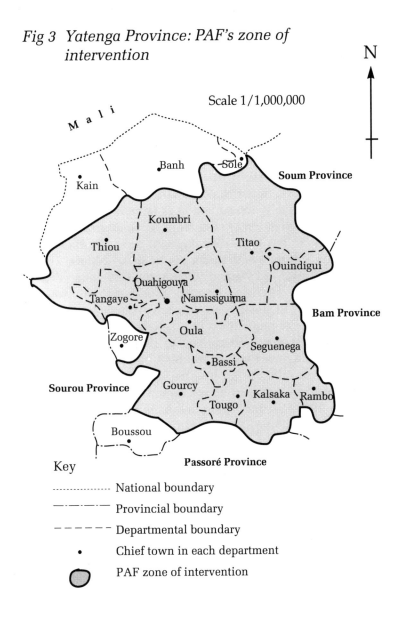

the famine of the 1970s, Oxfam became increasingly aware of the importance of finding lasting solutions to the causes of the famine. The search for solutions focused on the need to restore vegetation cover in the Sahel region, because deforestation had been identified as the most important cause of soil degradation, drought, and crop failure. Unnoticed or ignored by respective governments, the process of deforestation had been gathering pace several years before the famine.

As early as 1917, French colonial authorities had noted the process of degradation of natural resources. However, colonial policies, rather than slowing down environmental degradation, actually intensified it, as farmers were obliged to increase the cultivation of cash crops, such as groundnuts, to raise income to pay colonially imposed taxes. As the Rassam Naaba (Minister for Youth under the traditional Mossi system of administration) explained, 'Before, people were growing food crops. The colonisers came to impose cash crops on us. Every household head was required to bring a specified number of tins of groundnuts or a certain quantity of cotton. Sometimes farmers were obliged to suspend growing food crops, to avoid running into trouble.' Increasing population pressures exacerbated the problem, and led to a situation where about 11 per cent of the region's land was unproductive by 1973. Yet the region had once been prosperous, as Boureima Sawadogo (quoted at the head of this chapter) recalls.

The vegetation of Yatenga

The vegetation of the province is characterised by sparsely populated trees (mainly thorny trees), shrubs, and grasses. Trees which have survived are essentially drought-resistant species such as the sheanut tree (*tanga* in Mooré) — *Butyrospermum parkii*; the baobab tree (*toega*) — *Adansonia digitata*; the kapok (*voaga*) — *Combretum micrantum*; the tamarind (*pusga*) — *Tamarindus indica*; the néré (*roanga*) — *Parkia biglobosa*; and the acacia species such as mimosa (*zaaga*) — *Acacia albida*. It is from these species that households obtain their construction timber and fuelwood.

Existing tree species also serve as important sources of food during the seasonal periods of food shortage. For example, the sheanut fruit is a source of food when its outer cover is eaten, and provides much-needed oil when its seed is processed into butter and added to household meals. Also, baobab leaves are important as vegetables in the preparation of sauce normally eaten with the main cereal meal, *tô* (thick porridge). Néré (*roanga*) seeds provide a source of protein when their seeds are fermented. As Fatimata of Ranawa explains, 'The tree plays a vital role in our daily life. We use its wood to make stools to sit on and build our houses, and we make door frames with its branches. We relax under its shade when it is hot, especially when we are out in the fields. Its leaves and fruits provide us with food. My son, Hamidou, is always out roaming the bushes in search of fruit during the hunger season.'

Despite the vital role that trees play in the life of the people of Yatenga, it is evident that shrubs are replacing trees as increasing pressure on forest resources lays waste what was once a forested area. Ninety eight per cent of the province's energy is derived from forest resources, with consumption of wood energy estimated to have been about 361,500 tonnes in 1986. Despite this heavy reliance on fuelwood, the use of improved woodstoves is not widespread enough to make any impact on the rate of fuelwood extraction from the fast-disappearing forest resources.

Relief and soil types

The loss of vegetal cover is only one of the province's problems. Yatenga is generally a flat area. With the exception of a few hills (averaging between 400 and 600 metres above sea level), predominantly around Ninigui in the district of Koumbri, the landscape of Yatenga is essentially a plateau. This landscape, combined with the absence of tree cover, has created favourable conditions for soil erosion.

The rate of soil degradation has been so rapid that, with increasing sandstorms, the area risks being transformed into a desert. Local farmers recognise the declining fertility of their

Fig 4 Yatenga Province: relief

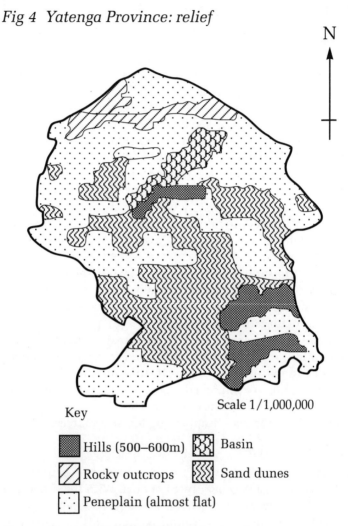

Key

- ▨ Hills (500–600m)
- ▨ Rocky outcrops
- ⋯ Peneplain (almost flat)
- ▨ Basin
- ▨ Sand dunes

Scale 1/1,000,000

(Source: First Five Year Plan for Popular Development in Yatenga Province)

fields when they regularly feel the texture of the soil or taste it during the farming season in order to assess its nutrient content. As they are always quick to point out, 'These days, our soil no longer has any taste.'

Fig 5 Yatenga Province: vegetation

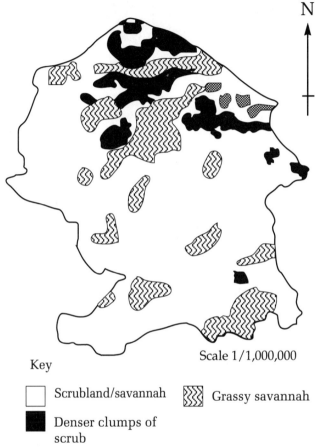

Key: Scrubland/savannah; Grassy savannah; Denser clumps of scrub

Scale 1/1,000,000

(Source: First Five Year Plan for Popular Development in Yatenga Province)

There are three major soil types in Yatenga. There is what is known as the *zecca*, gravel-like soil composed of rock debris 0.5 cm to 5 cm thick, on which farmers grow local beans, groundnuts, and late millet. Although its stony features protect the soil against erosion during heavy downpours of rain, it has a limited water-retention capacity.

Gully erosion, Yatenga Province

This usually leads to a rapid loss of moisture during short periods of drought, thereby threatening plant growth. This soil type is most common in the villages of Ranawa, Goumba, and Magrougou. Although *zecca* is averagely fertile, local farmers say that its major limitation is its poor water-retention capacity, in an area of declining rainfall.

There are also reddish-brown (ferruginous) sandy-textured soils known as *bissiga*. Although such soils have less than average soil fertility, they have a higher water-retention capacity. The hard rock beneath these soils allows limited water infiltration. As a result, its silt-laden surface easily gets washed away when there is torrential rain. Moreover, when there is a continuous downpour, such soils easily get flooded and can retard plant growth. *Bissiga*, however, has an advantage in being relatively easier to till. The main crops grown on such soils, most commonly found in the villages of Tanghin Baongo, Noogo, Ninigui, and Sologom-Nooré, are the 80-day maturing sorghum, millet, Bambara nut, groundnuts, and cowpea.

A third type of soil is the alluvial or clayey soils, known

locally as *bagtanga*, which are mostly found in low-lying areas. 'Black soil', as local farmers refer to it, is fertile, but rather difficult to work on because of its sticky or clayey characteristics, which slow down the rate of cultivation, especially when using the hand-held hoe. These soils are nevertheless eagerly cultivated, because of their higher plant-nutrient content and better water-retention capacity. Such soils are suitable for rice cultivation, as well as for the 120-day maturing millet, cotton, maize, sesame, and okra. Their disadvantage lies in the ease with which they get flooded when the early rains are heavy and prolonged, conditions which can hinder early planting. *Bagtanga* can be found throughout the province where there are patches of low-lying areas.

Apart from these soil types which have some agricultural use, there are often barren patches of land, known locally as *rassemponiga*. Local women rely on such land to dry leaves (as a way of preserving them), to be used later in the off-season period in the preparation of sauce.

The farming system

It is on the soils described above that the majority of the people of Yatenga have to eke out a living, mainly through the cultivation of sorghum and millet. Of the 369,000 hectares of cultivable land (about 30 per cent of total land area available), about 27 per cent (98,275 ha) is estimated to be actually under cultivation. The remainder has lost its fertility and is of little value as farmland. About 80 per cent of this cultivable land is under intense permanent farming, within the framework of what can be described as an extended-family farming system. Each family has access to about 3.6 ha of land, with which it is expected to produce, using a labour force of (on average) four people, the food needed to feed (on average) eight people. On this land each family has to produce its sorghum and millet requirements; the average yields are estimated to be 453 kg/ha and 423 kg/ha respectively. The main farming implement is the traditional hoe. Mossi women have skilfully mastered its use,

Mossi women, adapting to new roles in recent years, have skilfully mastered the use of the traditional hoe

compelled (by high male labour migration and the difficulty of producing enough food) to work side by side with their men. Traditionally, Mossi women were not supposed to participate in the weeding of millet and sorghum (the bulk of the farm work), although they were often responsible for sowing and harvesting of the crop.

Local farmers face an up-hill task to increase yields. In the first place, there is no shifting cultivation, since there is no available fallow land; so the soil has no opportunity to recover its fertility. Secondly, intensive farming is further handicapped by the lack of opportunities for directly increasing the fertility of existing farmland. Farmers rely mainly on organic matter, consisting of a mixture of animal dung and household refuse, which decomposes very slowly. The use of chemical fertilisers is limited not only by high costs but also by local farmers' justified conviction that chemical fertiliser dries up the soil. In an area of declining and unpredictable rainfall, and lack of access to irrigation, chemical fertiliser does not seem to be a realistic option.

Yatenga: A battle for survival

Water resources

The province has two watercourse systems: the north-west Niger river basin, consisting of small tributaries of the Niger river, which dry up on the approach of the dry season, and the main watercourse system, the southern Nakambé river system which forms part of the Nakambé basin (formerly known as the Volta river basin), whose tributaries retain some water for a slightly longer period. In addition, there are pools and ponds scattered around the south of the province, and some basins in the north.

Apart from a tributary of the Nakambé river (formerly Black Volta) which crosses the villages of Longa and Goumba in the district of Namissiguima to the east of Ouahigouya and usually contains some water, the only natural water sources are ponds and pools in low-lying areas. Although such natural water reservoirs used to be an important source of water for livestock, they now tend to dry up shortly after the rainy season ends.

The annual average rainfall stands at between 650 and 750 mm. The highest ever recorded was 1,020 mm in 1922, and the lowest was 358 mm in 1983. Between 1968 and 1987, average annual rainfall in the province declined by 200 mm

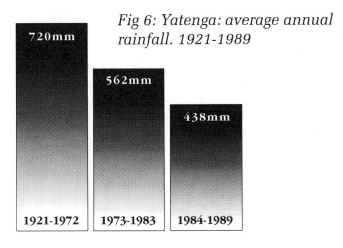

Fig 6: Yatenga: average annual rainfall. 1921-1989

(20 per cent). According to the Regional Centre for Agro-Pastoral Development (CRPA), an annual average of 513 mm of rain was recorded between 1981 and 1990.

Apart from generally declining rainfall levels, the dry season now seems to last longer. The number of rainy days recorded in 1979 was 64. In 1983, however, it fell to only 48. Furthermore, there is usually a dry period lasting 10 to 25 days at some point in the sowing season. This dry period often causes irreparable damage to the main millet crop. In addition to the short dry period within the rainy season, there are times when a significant proportion of the annual rainfall is concentrated within a few days. In 1957, for example, one quarter of the annual rainfall (about 179 mm) fell on one day. It is this situation which makes the *pattern* of rainfall precipitation far more important to farmers than the total amount of rainfall per season.

The declining rainfall pattern is also reflected in decreasing water availability, with the water table now so low that households can no longer easily dig traditional wells as a source of water for household use. Ground water can be found around three water tables: the Cambrian table at Titao, Tiou, and Kalsaka, the Cambrian granite table at the extreme west of the province, and the pre-Cambrian table in the north-east. In the village of Longa, the water table is estimated to be now 20 to 25 metres deep. At Mogom in the south, it is ten metres deep, while at Ranawa in the south-west it averages 16 metres. Such levels are virtually impossible to reach with traditional implements for digging. The declining water table affects pastoralists within the province hardest, because the absence of pools and ponds compels them to shift their migratory patterns farther south in search of water and pasture for their livestock. The movement of pastoralists is determined primarily by the changing seasons.

Seasonality

The rainfall pattern discussed above is the most distinguishing feature of the seasons of the province. The farming system is defined by two seasons, the rainy season and the dry season. The dry season, known locally as *sipalgo*, is divided into two parts. The first, usually cold, begins in November and ends in February. Cold weather increases progressively, reaching its climax in January. The weather begins to warm up only in February. During the cold period, households usually light a fire outside the homestead in the mornings, and group around it to keep themselves warm before sunshine warms up the air. Groundnuts are roasted in this fire to provide people with the energy to start the day. At night, slow-burning wood is lit to warm up rooms, especially when there is a new-born child.

As the second part of the dry season begins in March, ponds, pools, and wells begin to dry up. Trees shed their leaves and begin to dry up. Vegetation easily catches fire as green leaves turn dry and the *harmattan* gathers momentum. The *harmattan* is a cold abrasive wind that sweeps across the West African sub-region from the desert in the North, blowing southwards towards the coast. It carries along with it a lot of dust and sand. In its most severe form, it blocks out the

Approaching sandstorm near Titao village

sunlight, the sky darkens, and visibility is reduced to about 100 metres. To protect themselves from this abrasive wind and sandstorms, local people often wear headgear which entirely covers the head, except for the eyes. In this climate, very little agricultural activity takes place: the land becomes sunbaked and begins to crack in places, as temperatures rise to 40°C. The hottest month is April, when temperatures can sometimes reach 45°C. The arrival of the dry season coincides with a decline in grain stock, with household granaries often half-empty. Household heads begin to ration out cereal grain in the hope of making it last till the farming season. During this lean period, known as the 'hunger period' or *la soudure* (the French word for 'welding'), farmers can only wait anxiously for the rainy season to bring better times.

Signalled by the appearance of new leaves and flowers on certain trees in May, the rainy season proper, known as *seongo* in Mooré, starts in June and ends in October. As farmers wait anxiously for the rains, they begin to prepare their fields. When the first rains arrive, the landscape comes back to life. As grasses begin to grow, what once looked barren and desolate turns green and lush, and local people are full of hope. Iddrissa from Gourcy comments: 'If every time could be like the rainy season, we would not complain.'

The rainfall distribution is uneven and unpredictable, with short periods of drought mixed with periods of flooding. The life of the entire community is determined by the pattern of rainfall. The rainy season is the period which determines if a household will produce enough food to last through the dry season into the next farming season. For many households, it is a make-or-break period, as they deploy all their labour and resources to produce much-needed grain. But if the rains fail, farmers have limited alternatives. Although tools such as hoes and pots are produced locally, handicraft production is not a major source of cash income. Similarly, although women do process sheanut fruit into butter for sale, and despite the increasing importance of dry-season market gardening, the economy of the Yatenga household is still heavily dependent on rainfall for the cultivation of sorghum and millet.

Socio-economic infrastructure
Roads

Yatenga Province has 'first class' and 'second class' roads. Both types are generally untarred, irrespective of their class, and together stretch for 359 km, linking the province with its neighbours. There are also inter-District feeder roads which are badly maintained, as well as village paths that link villages to Ouahigouya. In all, there are about 800 km of road; some motorable all year round, others seasonal (unusable during the rainy season because of mud), and the rest unmotorable unless they are regraded. On the whole, the road network of the province can be described as rudimentary.

Health services

Table 1.1 (overleaf), on paper at least, suggests that health services in Yatenga are adequate. In reality, while the figures for basic health services might look impressive, most of the infrastructure (such as the primary health posts) is no longer in use. Some of the buildings have started to collapse, as the momentum which led to their construction was lost following the assassination of President Thomas Sankara in 1987. Within 15 months of Sankara's rise to power in the Revolution of 1983, primary schools, clinics and maternity centres, dispensaries, retail shops, dams, reservoirs, wells, and boreholes were constructed (relying mainly on mass mobilisation of labour) nationwide. The mobilisation effort was frequently presented in military language — 'Operation Alpha Commando' (the literacy campaign), 'The Rail Battle' (the attempt to link Ouagadougou with the manganese-rich site of Tambao in the north), 'The Battle of Kompienga' (the hydro-electric project) — and implemented with military-style haste.

Yatenga, like many other provinces, benefited from the rapid expansion in socio-economic infrastructure. The extended programme of immunisation aimed at protecting children against killer diseases such as yellow fever, polio, measles, diphtheria, and tuberculosis did help to reduce the

Table 1.1: Health infrastructure in Yatenga Province

Facility	Number
Central regional hospital	1
Medical centres	5
Health and social welfare posts	22
Dispensaries	16
Anti-leprosy and tuberculosis centre	1
Regional drug warehouse	1
Drug stores	36
Primary health posts	612
Maternity clinics	3
Centre for nutritional recovery	4
Social welfare centre	1
Day nurseries	2

(Source: Plan d'action de la province de Yatenga, 1991, Ministère du Plan)

child-mortality rates. Also, four diet-improvement centres were built at Ouahigouya, Titao, Séguénéga, and Gourcy to help severely malnourished individuals to recover. The one at Ouahigouya has 20 beds for taking in malnourished children while they regain their strength. This initiative is, however, undermined by widespread malnourishment in the province. There are also two day nurseries in Ouahigouya, which are meant to make it easier for working mothers to go out to work.

Although the goal of extending basic rural socio-economic infrastructure during the Sankara era was a laudable one, it is debatable whether there was a clear assessment of the implications of such an expansion. The overall reliance on popular enthusiasm ignored the practical issue of whether or not such endeavours were sustainable. The failure of the State to ensure that these structures were truly operational probably served to dampen the hopes of the people for improved living conditions. For example, local women who were trained as midwives, on the assumption that the local communities themselves would mobilise resources to pay for their services, soon realised that they had to do it on a virtually voluntary basis. They rapidly lost interest as they became preoccupied with their own struggle for survival.

Health infrastructure in the province in reality is grossly inadequate. There are four doctors for a population of 586,000, the equivalent of about 146,000 people per doctor. These four doctors are supported by 11 health assistants, six State-registered nurses, 26 qualified nurses, six trained midwives, and seven traditional birth attendants. This limited health coverage means that most patients have to rely on a combination of modern and traditional medicine for their health care. Modern medicine is particularly expensive in Burkina Faso, where every drug has to be paid for by patients themselves, in a currency which local people have difficulty obtaining. Unlike traditional medicine, fo which payment is usually made with livestock (although cash payment is becoming widespread), local farmers usually have to sell off some livestock in order to obtain the cash needed to pay for modern medicine.

To make matters worse, the recent agreement with the International Monetary Fund for a structural adjustment loan requires that hospitals be freed from State control. This means that the full cost of modern medicine has to be borne by patients. However, as the Provincial Medical Officer for Health in Ouahigouya points out, 'The people already have a very low purchasing power. If those who are sick even now cannot find the money to buy the medicine prescribed for them, what will happen if they have to pay even 300 FCFA as consultancy fees or for getting a boil removed when the hospitals are privatised?' Even when they find the money, they are frequently unable to get the treatment they need, either because local hospitals do not have the necessary equipment, or because the drugs prescribed are not available in local pharmacies.

It is this situation which forces local people in need of treatment to combine both traditional and modern medicine. So difficult is the access to modern medicine that many people visit the hospitals only when their traditional medicine fails them. Yet their attitude to modern medicine is conditioned not only by cost, as Kaboré of Gourcy explains: 'There are certain diseases that the white man's medicine can never cure. When you catch this illness, you know that it is no use going to the hospital.'

Table 1.2: Distribution of households according to type of medical care

Type of medical care used	No.	Per cent
Traditional medicine	143	12
Western medicine	214	18
A combination of the two	852	70

(Source: Results of questionnaires)

Domestic water supplies

The problems caused by Yatenga's inadequate health-service coverage are compounded by limited access to good drinking water. Traditional wells constitute the main source of water for many households. Such wells are hand-dug to a depth of about 15 m and a diameter of 0.8 m. Because of the declining level of the water table, wells have to reach even deeper to be able to provide water, a depth which makes the drawing of water a tiring job for women (who in traditional society are responsible for providing the household with its water requirements). In 1990, there were only 889 permanent water sources in Yatenga, out of which 460 were functioning bore-holes providing good drinking water, according to the Water Ministry. Although there were 7,552 traditional wells, only 429 had water in them all year round. Many traditional wells dry up by December, not long after the farming season has ended. Despite this situation, about 65 per cent of households rely on traditional wells for their water requirements, as shown by Table 1.3.

Table 1.3: Distribution of households according to source of water

Source of household water	No.	Per cent
Bore-hole	415	34
Traditional well	783	65
Stream	11	1

(Source: Results of questionnaires)

With 65 per cent of the people relying on traditional wells, most of which run out of water as the dry season advances, finding and collecting good drinking water is a task which takes up a lot of the time of the Yatenga woman. Although the government's five-year development plan had aimed at providing each person with about two buckets (20 litres) of water each day, this represents only about two-fifths of daily household requirements for drinking, washing, and cooking — not to mention the needs of livestock. In reality, most rural households do not obtain even the minimum requirement. The gravity of the situation is illustrated by a comparison with the rate of water use in Europe, where one flush of the toilet, for example, consumes about eight litres of water.

However, the problem is more than just the lack of water. The *quality* of water consumed has increased the health risks to the people of the area, especially since these wells are frequently polluted by dust storms. Not surprisingly, diseases such as diarrhoea and infestation by guinea worm are common in the province. Other diseases of poverty, such as tuberculosis and leprosy, are also prevalent. However, a campaign against leprosy has resulted in a huge drop in the number of sufferers, from 1,019 in 1980 to only 279 in 1990. According to the Ministry of Health, the aim is to eradicate leprosy by 1994.

Educational provision

During the 1989/90 school year, Yatenga province had 222 primary schools, with a total of 635 classes and an enrolment of 36,151 pupils, about 32 per cent of children of school-going age. Although this is higher than the national average of 29 per cent for the same period, according to the Ministry for Primary Education and Mass Literacy, it is still inadequate, considering the high population density of the province. Furthermore, while school-attendance rates may seem high, adult literacy rates are low.

Table 1.4: School-attendance rates (7-to-14 year age group)

Year	Yatenga Province Percentage	Burkina Faso as a whole Percentage
Before 1980	8	12
1986	17.3	20.4
1990	24.8	27

(Source: Plan quinquenal,1986-1990, Ministère du Plan)

A good measure of rates of adult literacy is the level of female literacy, especially as there is invariably a higher ratio of women to men, partly as a result of high rates of male migration from the province. According to the 1985 national census, women constituted 54 per cent of Yatenga's population. Yet according to our study, nearly 90 per cent of the women in the area have never been to school. Those who have had any literacy training at all have been to Koranic schools, with an insignificant number (1 per cent) pursuing adult literacy classes. The adult education scheme teaches local people how to read and write in the local dialects. It also includes basic arithmetic, elementary hygiene, and management. The data in Table 1.5 are derived from women's responses to questions on the type and level of education they have received. The results suggest extremely low adult literacy levels.

Table 1.5: Educational levels of women in Yatenga

Level	No.	Per cent
Never been to school	1073	89
Rural/Koranic school	33	3
Adult literacy classes	16	1
Primary school	4	0.3
Secondary school	3	0.2

(Source: Results of questionnaires)

There were, as of 1989/90, five secondary schools running 57 classes with 3,160 students. There were also three private schools with 15 classes and 792 students. In addition, the province had a girls' technical institute with 88 students. However, with the exception of the districts of Gourcy, Séguénéga, and Titao, which each have a training college, these secondary schools are concentrated mainly in the provincial capital, Ouahigouya.

The co-ordinator of higher-level education in the province argues that there is a general awareness of the benefit of education within the province, and that all parents would like their children to be educated. Despite this, parents are often compelled to withdraw their children from school owing to their inability to pay — as happened, for example, in the 1990/91 academic year, in the district of Séguénéga. Although the enrolment fee in 1989/90 was 18,000 FCFA for new pupils and 13,000 FCFA for continuing students, many parents find the fees too high. As a result, some heads of institutions avoid insisting too much on payment, for fear of losing most of the pupils.

Migration from Yatenga

Low levels of education have certain practical implications for households within the province. Apart from migrating as labour to neighbouring countries, there are very few options for obtaining wage incomes internally. Burkina Faso is a country where the State is the single largest employer. Previously, an education was a way of obtaining a salaried job, and Burkina Faso's workers had successfully secured, through their relatively powerful trade union movement, higher wage levels than those in some neighbouring countries such as Mali or Ghana. Now, even advanced qualifications are no guarantee of a secure job. This means that most people in the region are restricted to working the land. It is the search for cash income and a declining ability to feed off the land which have been responsible for the high rate of migration to Côte d'Ivoire or Ghana, mainly as labour on cocoa and coffee plantations. With declining opportunities in Côte d'Ivoire, many young men are drifting to the cities or towards the south-west of Burkina.

Harouna Ouédraogo, secretary to the Ranawa Village Group, observes: 'Almost everybody in this village has left for Côte d'Ivoire. Some have also gone towards Bobo-Dioulasso in the south-west in search of more fertile cultivable lands. But we stay here. It is better to remain at home. The problems facing those who emigrate to Côte d'Ivoire are as serious as those that caused their departure in the first place. Indeed, there is no job for them other than the plantations. They depend on the plantation owner, and sometimes work the whole year round without wages. So they can't send any money to their relations back home; they can't even take care of themselves. They are no better off than if they had stayed here. Those who go looking for more fertile lands in other parts of this country are not better off, either. The journey is costly, since very often the entire family goes along. On arrival, they have to borrow food from the local people to start with, but it must be paid in two or three years if the crop is good. They can't hope to stand on their feet for at least three years.' Harouna believes that those who remain in Yatenga are better off. At least they are unlikely to incur huge debts, and can recover with relative ease if harvests are good. And with the introduction of *diguettes*, their hopes of self-sufficiency are higher.

Harouna has a point. With declining world prices for cocoa and coffee, the Ivorian economy has virtually gone into a free fall. Even government employees have had problems getting paid, while plantation owners are faced with a stock-pile of cocoa and coffee beans which they cannot sell. Also, while migrating south-west used to be a useful option, indigenous people of the south increasingly resent the rising concentration of Mossi migrant farmers in their midst. At first, they welcomed them; their culture required that they offer land, on a usufruct basis (conferring the right of use but not ownership), to any one in need of a plot of land to feed their family. However, as a young man from the south-west complained, 'The Mossi are now behaving as if they own the place. We have to let them know that this is our land.'

Yet young men still migrate, leaving their wives behind to bear the brunt of poor socio-economic provision and perennial food deficits. Iddrissa Compaoré is very clear in his mind why he has to migrate. 'We know nowhere is good these days. Yet it sometimes becomes a choice of how to die. The pain of seeing your family suffer, when there is very little you can do, is too much to bear. I am leaving, but not because I do not love my family. Whether I am around or not, they will have to manage without me. When I am away from home, at least I can bear any indignities just to earn an income. At home, your very honour is at stake.'

According to Haoua Ouédraogo of Goumba, the men who remain are not necessarily an asset. 'When the kids are hungry, they complain to us mothers. Their fathers often escape to the bush to avoid hearing complaints. We, the women, are always at home, and we have to bear the tears. Most often we have nothing else but the edible leaves that we pluck from the bush. We somehow manage to come up with something for the family. These leaves here can be eaten alone, or with a bit of flour if there is some millet available.'

Comment: the challenge facing NGOs

The situation described above poses enormous challenges for any organisation that sets itself an agenda to improve the living conditions of the people. Considering the scale of the problem, it is arguably a challenge which requires simultaneous action on all fronts. This is possible, perhaps, only under conditions of revolutionary socio-economic transformation. What makes the challenge for non-governmental organisations (NGOs) even more daunting is farmers' risk-avoidance strategies for survival, or what some describe as peasant conservatism. As Aly of Gourcy put it succinctly: 'Maybe I can take a risk on the national lottery, but I dare not gamble with what to grow on my field or how to grow. I wouldn't survive such a gamble.' As a result, local farmers are not keen to try out innovative ideas, until they have been proved effective beyond reasonable doubt.

Under the unfavourable conditions described in this chapter, it is worth asking why so many NGOs operate within the province. How can they tackle such a wide yet interrelated range of problems? In many projects, such as Oxfam's PAF, they bother because they believe that by prioritising their interventions, they can make a meaningful impact. Access to cultivable land, very limited non-agricultural sources of income, and the problem of deforestation are high on their agendas. The question of whether tackling those problems will make a significant impact on the living conditions of the people will be explored later.

2

External agency intervention in Yatenga

Why are so many non-governmental organisations (NGOs) concentrated in Yatenga Province? Valian Armadou, an official of BSONG, the NGO National Secretariat, suggests two reasons: the exceptional number of natural calamities which afflict the area, and the local people's reputation for hard work and willingness to collaborate on projects. However, Marcel Boureima, a worker with the Ministry of Commerce in the capital, Ouagadougou, takes a more cynical view. He alleges that external agencies often arrive unsolicited, but nevertheless welcomed, by village communities, promising to make a unique contribution to the development of the area. After a decade or so, when an agency's funding sources begin to dry up and it prepares to leave, villagers ask themselves what there is to show for the much-publicised intervention. Their living conditions have remained unchanged, and in certain cases have even deteriorated. As a Canadian NGO official working in the Nahouri province in the south-east of Burkina Faso once remarked: 'If we had distributed all the dollars we brought into the province among all the adults, perhaps we would have made a better impact.' Cynicism apart, it is undeniable that Yatenga is subject to one of the greatest concentrations of

NGO interventions in Burkina Faso. Furthermore, external agency intervention goes back more than three decades.

Learning from the past

External agency intervention in Yatenga, ostensibly to contribute to the development of the province, is certainly not recent, especially in the field of soil conservation. The European Society for the Restoration of Soil (known by its French acronym, GERES) had tried shortly after Burkina's independence, between 1962 and 1965, to control soil erosion in Yatenga. Their preferred manner of intervention was the construction of earthen dykes, using bulldozers and graders, across abandoned or degraded soils, as a way of preventing erosion caused by run-off water. Trenches 30 centimetres deep, with a gradient of 0.25 per cent, were dug parallel to the slope. Rain water was to be channelled through these trenches into outlets where stone-reinforced mud walls were erected to prevent further run-off. In all, about 120,000 hectares of land were 'treated', and 24 water reservoirs or micro-dams built.

GERES ended its intervention by 1966, but the effort to halt soil erosion in the province was taken up ten years later by a Rural Development Fund, funded by the World Bank and known by its French acronym, FDR, which was to some extent an improvement upon GERES. FDR sought to involve local Village Groups, unlike its predecessor, which saw no need to consult with or seek the advice of local farmers. Through these Village Groups, FDR hoped to encourage farmers to construct earth *diguettes* along contours determined by qualified topographers. *Diguettes* so constructed would then be reinforced, during the subsequent farming season, by planting local plant species such as *Andropogon gayanus*, known locally as *pitta*.

The people of Yatenga have mainly unpleasant memories of the attempts by GERES and FDR to halt soil erosion. Not only were earth dykes not maintained, but local farmers cut across them to retrace the footpaths which had once linked villages before becoming blocked by unsolicited dykes. Moreover, better-off sections of the local community lobbied to have their fields developed — to the detriment of poorer sections of society. According to Issoufou Ouédraogo, Chief of the village of Longa, who worked for GERES, 'The only positive thing about GERES was the water reservoirs they built, and the heaps of stone that they used to close off big gullies. Otherwise, the so-called ditches which were meant to halt erosion actually increased water run-off and intensified erosion.'

Some local farmers go even further, and question the scientific validity of the GERES approach. They ask what sort of agronomy these so-called experts had studied. As Ousmane Ouédraogo, a farmer of the village of Noogo, points out, 'Even if they are well maintained, earth *diguettes* are a complete menace, because they flood upland soils and dry up land below, turning it into rock. Stone *bunds* are better, because they are not easily damaged and require less maintenance. They also allow water to pass through, which enables you to grow crops in front as well as behind.' The Rassam Naaba (the Mossi Minister of Youth under the traditional system of administration), 60 years old and currently working for the department of animal husbandry, goes even further: 'They sent us fake engineers who came to destroy with their huge machines. They only came to destroy our trees and leave. They did not even ask the opinion of the local people.'

Recent NGO involvement

The influx of external agencies into Burkina Faso began after the drought of the 1970s. It attracted international attention to the problems of the region, and provided a good justification for operational involvement by NGOs. Before that, Catholic missionaries had the most pronounced NGO presence in the country. In addition to channelling relief aid through its relief wing CATHWEL (Catholic Relief Services), the Church was involved in promoting various forms of cooperative activity, mainly in the south-west of the country.

Table 2.1: Externally-funded organisations with offices in Yatenga

Agency	Areas of intervention
Oxfam (Agro-Forestry Project, PAF)	Soil and water conservation, agro-forestry, livestock, extension work, training, revolving grain stock
Association of French Volunteers for Progress (AFVP)	Soil and water conservation, micro-dam construction
Agro-Ecology Project (PAE)	Soil and water conservation, soil restoration, integrated agriculture/animal husbandry
'Six S' Movement	Soil and water conservation, collective farms, market gardening, animal husbandry, grinding mills, cereal banks, weaving, fruit and vegetable preservation, joinery, micro-dam construction
Desjardins International Development Agency (SDID)	Promotion and training for savings and credit co-operatives
North Yatenga Food Growing Project (PVNY) (para-statal)	Improving natural-resource management, supporting the marketing and supply of agricultural products and agro-industrial by-products, promoting women's income-generating activities.

External agency intervention in Yatenga

According to the NGO National Secretariat (BSONG), whose role is to co-ordinate NGO activities throughout the country, there are currently about 150 NGOs in Burkina Faso, contributing a substantial amount to development activity. The highest concentration of NGOs is in the Yatenga Province. According to BSONG, there are five NGOs (including Oxfam UK and Ireland, through its operational project, PAF) with permanent offices in Yatenga. Sixty others operate in the province, while ten more, not officially recognised by the State, operate there through individuals or other recognised institutions.

The most important externally-funded projects, in addition to PAF, are the Agro-Ecology Project (PAE), the 'Six S' movement, the French Association of Volunteers for Progress (AFVP), the Desjardins International Development Agency (SDID), and the North Yatenga Food Growing Project (PVNY), a para-statal organisation.

The Agro-Ecology Project

PAE started in 1981 with the German Programme for the Sahel. It was financed by the German Agro-Action Scheme and run by the German Voluntary Service Overseas. In 1991 it had six German staff and two Burkinabè. Its objective was to teach farmers about the judicious use of land resources, in such a way as to maintain ecological balance. Its philosophy is based on encouraging the voluntary participation of local farmers in its activities, without recourse to direct incentives. M. Gabriel Zida, Burkinabè counterpart manager in the project, comments: 'Some NGOs distribute food aid. We think that even if there is an emergency, you have to encourage people to rely on their own resources.' Nevertheless, PAE recognises the need to offer relief aid in time of catastrophe. In line with its stated philosophy, PAE prefers to lend, rather than give, its tools to farmers. As one of their agricultural engineers explained: 'The farmers are allowed to retain only the barest minimum necessary for the continuity of the work started.' PAE is involved in popularising the construction of *diguettes*, using the water-tube technique of contour determination pioneered by PAF. It is also involved in soil-improvement techniques like the construction of compost pits, and establishing tree nurseries for afforestation. This is usually accompanied by educating farmers on the process of soil degradation and ways of tackling it. Although it started with an expanded programme involving many villages, it has begun to scale down its activities. As Corinna Arnold, a staff member, explains: 'We realised that we were carrying out too many isolated projects. Henceforth, we will concentrate our efforts on a very limited number of villages. Six or seven will be the maximum.'

The Association of French Volunteers for Progress (AFVP)

AFVP is run by French agricultural volunteers who started in 1980 by constructing small dams in Titao, in the Yatenga Province. They also embarked on a four-year (1988-92) soil and water conservation programme in the same area, as well as in the neighbouring province of Bam. According to Omar Tapsoba, one of its animators, working in adjacent provinces enables them to make a comparison between results obtained in different provinces. The project's work involves educating and training farmers in soil and water conservation techniques. It has encountered certain difficulties, as a member of staff explains: 'In carrying out this work, we wanted to work with all the villagers. However, in certain villages, there are rival Village Groups. We ran into a serious problem of who to work with. What could we do? We tried to overcome this problem by by-passing Village Groups. Recently, we undertook a census to enable us to define production units with whom we can work. Only the future will tell us if this experience is conclusive.'

The Desjardins International Development Agency (SDID)

SDID is a Canadian NGO that started in the 1970s in the Bougouriba Province, in the south-west of Burkina. It mobilises cash and redistributes it in the form of loans to village co-operatives. As Marcellin Kaboré argues, 'If we estimate all the aid that this country has received, it is worth asking where it has all gone to. There is a need to mobilise our own resources.'

Desjardins was initially based in the south-west, but the Ministry of Rural Development asked it to widen its services to other provinces. As a result, it extended its operations to Yatenga and Kadiogo (Ouagadougou) in 1986. The results of the agency's work in Yatenga indicate some potential for mobilising savings. In 1991 co-operatives initiated by SDID managed to collect approximately US$335,000 in the form of savings, of which 21 per cent has been redistributed in the form of credit. This success enabled the agency to offer a grinding mill to a women's group in Bogoya as a reward for winning the first prize for efficiency in 1988; the profits from the mill have enabled the group to purchase another grinding mill. Credit of US$1,000 also enabled the Women's Group to start petty trading, which is proving to be viable. Yet it certainly has not been easy, as Kaboré explains: 'The urban savings schemes are quite advanced. In fact, they are the best organised. You know, people are generally very suspicious when it comes to mobilising savings. Moreover, we have established ourselves in a completely new area. Some NGOs have conditioned the people to receiving aid. We are against this, because it hinders people from taking responsibility for their own lives. We believe that it is necessary to begin from people's own resources in order to help them. You have to cultivate in people a more efficient management of their resources.'

North Yatenga Food Growing Project (PVNY)

PVNY is essentially a parastatal which operates more like an NGO. It was established with loans and grants from multinational agencies and a seven per cent contribution from the national government. It aims to promote decentralised, participatory, and multi-sectoral development by encouraging effective control by village communities themselves in the management of natural resources. In this process, its work focuses on promoting integrated agro-pastoral development by educating and training farmers, and supporting the marketing and supply of agricultural inputs and agro-industrial by-products.

The 'Six S' Movement

The 'Six S' Movement, and its network of Village Groups known as *Groupements Naam*, is an internationally known NGO with an area of operation which extends beyond Yatenga province and across neighbouring countries of the Sahel. The six S's stand for 'Se Servir de la Saison Sèche en Savane et au Sahel' ('Making use of the dry season in the Savanna and the Sahel'). Founded by Bernard Lédéa Ouédraogo, it developed, starting in 1967, out of his recognition of Village Groups' failure to act as motors of development in the countryside. Having worked with the State Regional Development Organisation (ORD), he was painfully aware that these institutions, inherited from French colonialism, had little relevance to the interests of rural people. As Lédéa points out, 'Villagers were baited with money, tools, food, etc. They worked as a group, but they lacked any conviction. Village Groups usually broke up as soon as they obtained inputs or credit.'

Yet Mossi traditional society did have its own indigenous forms of social organisation which, Lédéa observed, had enormous potential as basic units for rural development. The traditional Mossi social institution of *Kombi Naam* is a social and cultural organisation which renders service to its members during the farming season. In doing so, it also generates and saves income for the celebration of the *Naam* feast at the end of the farming year. So important was this feast (which partly functioned as an initiation into socio-cultural aspects of Mossi society) that young men (members of the Kombi Naam association) were forbidden to leave the village before the celebration. To do so would be the equivalent of desertion in military terms. Each young man took part in a role play, in which he adopted a young woman as his 'mother'. The challenge to young men was to dress up their 'mothers' adequately for the feast. To flee the village before the feast was tantamount to failing one of the tests of manhood. Lédéa Ouédraogo, as a Mossi himself, recognised the galvanising potential of the Kombi Naam association. Its main limitation, however, was that it ceased to function once the feast was over. Furthermore, the feast was not always a yearly affair, and was heavily dependent on the rural community's state of well-being, especially that of its farming system.

To overcome the essentially undemocratic character of Village Groups, he appropriated the essence of the Naam associations and sought to transform them into permanent features of village society. Groupements Naam, as they have come to be known, are the product of Lédéa's attempt at social engineering. They are now a permanent feature of Yatenga society, co-existing with Village Groups. However, Naam groups no longer hire out their services, nor do they hold celebrations. In their current form, they are essentially geared towards mobilising village communities to find solutions to their pressing needs. Lédéa Ouédraogo explains the Naam philosophy as follows: 'Our philosophy is to mobilise society based on what farmers know, on what they are, their experiences and wishes. In a nutshell, it aims to develop without destroying. It essentially helps local people to assume responsibility for their own development.'

The Groupements Naam received material, financial, and human resources from the International Association of Swiss Law which transformed the movement from the status of a local NGO into a transnational development agency known as 'Six S'. Since then, 'Six S' has developed a more diversified funding base, although its funding sources are probably not as reliable now as was the case in the 1980s. With activities that extend not only across Burkina Faso, but to other countries in the Sahel such as Niger and Mali, Groupements Naam are involved in the widest range of integrated development, embracing soil and water conservation, dam construction, dry-season gardening, cereal banks, adult education, women's income-generating activities, the manufacture of basic tools, and literally any productive activity which, with some external assistance, might improve the life of the rural community.

Groupements Naam and its founder, Lédéa Ouédraogo, have attracted wide international attention and publicity. According to Lédéa, their success and international acclaim have brought with them some problems. One is that 'I have been fiercely criticised by all those who dream of organising farmers on a "modern" model: intellectuals, staff of the State agricultural services, ORD [now CRPA], economists, and sometimes politicians. For me, however, the important thing is the farmers who follow me. There are today more than 4,000 Groupements Naam in the country.' However, despite the originality of Lédéa's initiative, these village institutions today bear little resemblance to their traditional forebears. In several villages in the Yatenga region, there is frequently little difference between active Village Groups and Groupements Naam. The use of the name 'Naam' often reflects more of an exercise in brand-differentiation than any substantial differences, with all the problems that maintaining brand-identity usually entails.

PAF: the early years
The origins

PAF was initially conceived by Bill Hereford, who was assistant Field Director for Oxfam (UK and Ireland) from 1976 to 1979 in Upper Volta (as Burkina Faso was then known). Hereford had been to the Negev desert in Israel on holiday, where he was struck by the use of soil and water conservation techniques for growing fruit trees. The relevance of the techniques was not lost on Hereford, working in a context of chronic and increasing deforestation. Determined to try out the techniques he had observed, he recruited Arlene Blade (a Peace Corps Volunteer) in 1979 to test the feasibility of the Negev techniques under local conditions in Burkina Faso.

Arlene Blade's mandate was to explore the possibility of

External agency intervention in Yatenga

Outside the PAF office in Ouahigouya

growing trees along *diguettes* constructed along the contours of the land. If the experiment was successful, it could become the basis for starting an afforestation project. In the meantime, it was basically a trial project under the title 'Projet Micro-Parcelles'. As part of her experiment, Arlene Blade organised farmers in eight villages to construct stone *bunds* on their fields and then plant imported trees along them. These trees were then left for observation for about 18 months, while she maintained regular contact with pilot farmers to discuss with them the progress of the experiment.

The first collaborators in the participatory research project were members of eight village groups (identified by a local forester for their enthusiasm for tree planting), or *Groupements Villageois*. The agenda at project-organised group meetings invariably focused on a discussion of the process of environmental degradation, and possible actions that Village Groups could undertake. Working through Village Groups was necessary for the management of *diguette* construction, for example in the organisation of stone gathering, and the conduct of participatory trials. It was also found that training local farmers to construct *diguettes* is best organised when they operate as groups.

The research was based on the hypothesis that halting and reversing the slide towards environmental catastrophe (especially deforestation) in a region which was experiencing declining annual rainfall had to begin with harvesting run-off water. Land degradation meant not only a loss of top soil, but also an increase in run-off, and a subsequent reduction of water infiltration. A second hypothesis of the research was that any technique aimed at halting the process of deforestation had to be an integral part of the process of households' struggle for survival. This would entail involving local farmers directly in the process of designing and testing techniques of water harvesting.

Although Arlene Blade did not stay long enough to follow her research to its logical conclusions, interesting results were already emerging from the experiment. Local farmers had found the stone *diguettes* useful, but their interest was not in growing trees. They had meanwhile been conducting their own experiment. Sorghum planted along the stone *diguettes* germinated and grew better than sorghum planted away from them. So when local farmers continued with their experimentation on *diguettes*, they were more interested in exploiting the technique for cultivating their basic cereals than in the possibilities for afforestation.

When Peter Wright took over from Arlene Blade in 1981 as co-ordinator of what had been renamed the Agro-Forestry Project (PAF), results from the experiment on stone *diguettes* already looked promising. A native of Yatenga, Mathieu Ouédraogo, had been recruited in 1980 as Arlene Blade's assistant. He and Wright now took the research into stone *diguettes* some stages further. Mathieu, the current project co-ordinator of PAF, remembers those days well: 'The objective was to undertake participatory research into soil and water conservation, which we hoped would produce results which would rely mainly on simple and inexpensive techniques for implementation. The idea was that if trials were successful, the techniques would be appropriated by agencies operating in the province, such as State agricultural services, or NGOs like the "Six S" movement.'

It was against this background that PAF continued with its research into appropriate soil and water conservation techniques. The project was conscious of earlier interventions in the province that had aimed at slowing down soil erosion. Two relevant lessons were learned from previous experiences, especially that of GERES. One was that capital-intensive techniques are not effective in halting the problem of soil erosion. The second lesson was that the involvement of local farmers in all stages of the process, from developing to implementing appropriate techniques, is indispensable to the success of any attempt to reverse the process of environmental degradation.

The process of adaptation

From the outset, PAF staff appreciated the need to learn lessons from other experiences which did not use capital-intensive techniques and to adapt their findings to local conditions. Lessons from the Negev showed that it was possible to reclaim degraded land and transform it into productive use such as fruit-tree cultivation, by using local materials to construct simple micro-catchments to harness run-off water. In Yatenga, however, this could not be done by merely replicating the techniques observed in Israel. It was important to find local parallels, based on which further developments could be made.

The indigenous farming system of Yatenga had also produced some anti-erosion techniques, but they had proved rather ineffective. For example, plant barriers have traditionally been constructed in flooded areas to hold back soil nutrients and prevent water run-off through gullies created by water erosion. There is also the practice of constructing barriers made of dead wood or millet stocks, and erecting them against the water current to hold back the flow of water. However, the speed with which plant material decomposes or is eaten up by termites has invariably reduced the utility of such techniques to an interim measure. Local farmers also sometimes grow a local grass species, *pitta*, perpendicularly along trenches or lowlands to combat water erosion. *(Pitta* had been particularly useful as a way of demarcating land boundaries, while its thatch served as raw material for roofing huts and making mats.) Similarly, the practice of leaving strips of grass at the bottom end of fields was another strategy for reducing soil erosion. The main handicap of grass barriers was and remains their dependence on adequate soil moisture, which is determined by the levels of annual rainfall. As a result, they have tended to shrivel and die during the dry season.

Although the technique of stone terracing was practised among the Mossi, it was comparatively underdeveloped and actually in decline. The most serious handicaps were the labour and time needed to gather stones (in an area where many members of the active labour force customarily migrate in search of paid work), and its ineffectiveness in holding back water over large tracts of land. Water run-off always found a way of by-passing these stone lines.

It was in this context that Oxfam's PAF continued its development and adaptation of soil and water conservation techniques. The research team identified several variables which could determine the outcome of the technique. First, there were organisational problems: how to organise labour to collect stones in sufficient quantities to construct *diguettes* across fields. Linked to this was the question of how to motivate farmers to undertake the exercise on a wide scale. The technical challenge was to find a way to determine contour lines, in order to harvest run-off water. The most critical aspect, however, was how to develop a simple instrument to do this job, using local resources.

Why Yatenga?

The choice of Yatenga as a test region had been made on the grounds that the area offered ideal conditions for trying out any techniques of soil and water conservation suitable for semi-arid environments. Its traditional anti-erosion measures had a potential for further development. The area was also undergoing rapid deforestation, particularly marked since the drought of the 1970s. Furthermore, as we saw in the preceding chapter, it is an area of high population density and diminishing access to cultivable land. With only 27 per cent of total surface area useful as farmland, every inch of land that can be recovered is worth recovering. Yet there was some room for manoeuvre, in contrast to the Sahel Province of Burkina Faso, which is virtually covered by sand dunes and where rainfall levels are even lower. Yatenga combined both aspects, with sand dunes in its north-eastern districts (where pastoralism is the main agricultural activity) and semi-arid eco-systems towards the south and south-west.

As a visitor to Yatenga once remarked, 'A tour of the province leaves you with a strange mixture of hope and despair. In the dry season, especially towards the north-east, the area looks like a hopeless case. As you move south and towards the south-west, large clusters of tree cover suggest that something can be done. When you visit the area in the rainy season, you are filled with optimism, as the landscape turns green and lush.' The search for appropriate water-harvesting techniques aimed to transform despair into hope, and build upon hope to give the farming community a more promising future. The challenge was to find simple yet effective instruments which could help farmers to determine contour levels on their farms — for without accurate contour determination, *diguettes* would have very little effect.

The water tube

A breakthrough was made by Peter Wright and his team when a simple instrument was developed to assist farmers in determining contour lines — lines which define positions of equal height above sea level. This is necessary, if barriers are to be built to prevent rain water from flowing from higher to lower levels on farmland. The instrument which was to become central to *diguette* construction is known as the water tube, or water level. The idea was borrowed from an organisation called CIEPAC, based in Dakar (Senegal), through Pierre Martin of IPD/AOS in Ouagadougou.

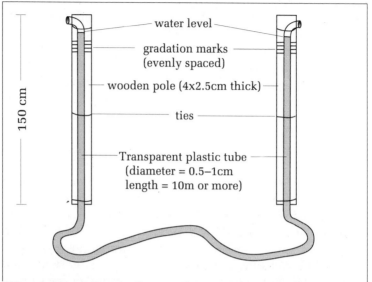

Fig 7: The water tube (source: Rochette (1989))

The water tube consists of two equal poles of wood, measuring about one and a half metres long and 4 cm x 2.5 cm wide. Both poles are graduated to show equal levels. This is done by placing the poles side by side on a flat and even surface, and marking out equal distances on both planks. For example, both poles can initially be divided into five equal distances, using a pen or charcoal (for demonstration purposes). These distances are in turn sub-divided into smaller equal units. This allows for greater precision in determining equal heights above sea level.

A transparent plastic tube, 10 metres or more long, and with an internal diameter measuring between 0.5 cm and 1 cm, is then tied to the two poles with a piece of rubberised cord (made for example from a disused bicycle inner-tube) at three places, with both ends of the tube turned upwards. This apparatus is then filled with water (ensuring that no air bubbles are trapped inside). Two people, each holding one of the poles, then pace the ground, observing the levels of water in the tube, and calling out their respective levels to each

External agency intervention in Yatenga

A wooden pole with gradation marks and a transparent plastic tube attached: one half of the PAF water tube

Behind the Lines of Stone

'To determine your contours, you place one pole at one point (held by your partner), and you walk with the other pole to a position on your field which your eye suggests could be level with the first pole. You read off your level and shout your measurement to your partner.' (Ousseni Ouédraogo)

other, until they are standing in positions where the water levels are identical: then they know that they are both standing on the same contour line. The line can then be marked on the ground with sticks.

The adaptation of this basic instrument for use in *diguette* construction can be considered revolutionary, for several reasons. Although it was basic and added nothing new to scientific theory, it was to become an important element in the battle to harness run-off water. In the first place, it reduced or eliminated the need for any professional topographers to help farmers to determine contour lines. In a poor country like Burkina Faso, where formal educational levels are low, any dependence on trained topographers would invariably mean the employment of paid workers. In a situation where government has limited resources, it can be argued that any resources available for the payment of trained topographers could be better spent elsewhere. Moreover, the use of professional topographers would take the initiative out of farmers' hands. The development of the water tube made it possible for local farmers to be involved directly in the struggle to halt the process of soil erosion. For once, they would not be reduced merely to a labour force carrying stones, or observers of the 'ingenuity' of 'expert' technicians, but active participants in the process of innovation. Thirdly, the materials needed to produce this instrument are so basic that it does not require any significant importation of foreign inputs. It could therefore be made accessible to poorer households at an affordable cost.

The water tube therefore solved the problems that had undermined the effectiveness of the traditional technique of stone lining. If trials proved to farmers that their traditional technique was actually effective in increasing crop yields, it could encourage them to undertake a creative adaptation of other indigenous strategies within the farming system. Yet the above technique had to be tried out over time on farmers' fields — not only to convince the local community and agencies operating in the region of the utility of the technique, but to identify the technical and organisational problems that the new instrument would produce.

From experimental exercise to operational reality

The research phase began in 1979 and ended in 1982. The problem that now confronted the team was what to do with the results of their research, as Oxfam had not taken a decision to become operational in the Yatenga province. Furthermore, the original intention had been to transfer the results to other agencies. Mathieu Ouédraogo explains why the transition was made from a research exercise to an operational project: 'The ORD [Regional Development Organisation, now CRPA] were not convinced of the efficacy of the technique. They pointed out that even if they were convinced, they lacked the resources to carry out effective extension work on the new technique.' There was, however, the home-grown NGO known as the 'Six S' movement.

According to Mathieu, they were clearly interested in the new technique of soil and water conservation. Together with Peter Wright, then the co-ordinator of PAF at Ouahigouya, they even began training local animators of the movement in the new technique.

PAF was, however, confronted with a dilemma when it tried to transfer the technique to other agencies. On the one hand, the State agricultural agency was clearly not enthused by lining up stones using an improvised water level, rather than the standard spirit level. On the other, the 'Six S' movement began to envisage an operational scale which clearly conflicted with the pioneers' goal of a modest intervention. According to Mathieu, 'Six S' drew up an initial budget which incorporated reafforestation, building animal enclosures, and buying carts, barrel containers, and watering cans as well as paying for animators.

The anxiety concerned not the need for such items but whether, as a first step, it was appropriate to begin on such a scale. High-profile initial interventions with a heavy input of external resources usually have the effect of reinforcing external dependence and killing off local creativity. The PAF team preferred a participatory approach, which would identify solutions to problems as they emerged during implementation, rather than anticipating problems and prescribing solutions in advance. This meant that the project had to start modestly. It was as a result of these realities that the project decided to become operational, for an initial period of three years.

The first fully operational phase (1983 to 1986) was to a large extent a continuation of activities undertaken during the research phase. Training was given directly to local farmers and extension agents of the ORD who wanted to participate. These people were trained in the use of the water tube. After each two-day or three-day training session, each Village Group was given a water tube with which to experiment on the members' own fields. Inter-group visits encouraged observations and the sharing of ideas among farmers. The purpose was to reinforce the principle of creative adaptability.

External agency intervention in Yatenga

Farmers using a model to demonstrate the increased capacity for water-retention when soil is treated with stone diguettes

Results of the first operational phase

By 1984 about 500 farmers had been trained in more than 100 villages of the Yatenga province. The project conducted a survey of 313 fields on which farmers had used the water tube to construct *diguettes* on their fields. The survey showed that farmers constructed barriers 10 cm to 50 cm high and 10 m to 100 m long, along contours determined with the water tube. *Diguettes* were constructed on individual fields, rather than on common land such as a watershed. They were mostly of permeable design (using rock, stalks, or branches tied in bundles, or live vegetation), allowing run-off water to pass through them. It was rare to see any earth *diguettes*, because farmers preferred rocks, often eagerly transporting them with donkey carts from distances up to 4 km.

The survey showed that 70 per cent of fields treated were bush farms. Bush farms differ from farms around the homestead because of their lower soil-nutrient content. Farms around the homestead are frequently fertilised with animal manure or human waste. Around Ouahigouya, 45 per cent of fields treated had been abandoned, because of their land-barren, crusted soils. *Diguettes* constructed on these fields succeeded in capturing water run-off (usually rich in soil nutrients) from degraded surfaces farther up the slope. On one third of the fields surveyed, farmers made a point of trying to divert waterways into their fields. On about 58 per cent of fields treated, *diguette* construction was the main technique of soil treatment used. The only addition was the application of another traditional technique on 33 per cent of fields surveyed.

The traditional technique of *zay* consists of digging holes all over the field as a method of harvesting water into pockets, as well as concentrating a mixture of manure and other organic matter on the areas where it is most needed. *Zay* had been a preferred technique of harvesting rain water and channelling it directly to nourish cereal crops. It was and remains a practice which plays a key role in helping households to harvest a crop, even if a modest one. Digging *zay* holes all over the field is undoubtedly a laborious exercise, given the nature of the soils and type of implement (the traditional hoe) used. Yet for local farmers, despite the tediousness of the exercise, these holes hold a key to their survival.

Comment: some lessons from the early years
Aim low

The early years of PAF contain some important lessons for NGO intervention. In the first place, a modest intervention with limited resources has the initial advantage of attracting mainly poorer households. This can be useful for designing techniques which will be accessible to those with limited resources. Better-off households in search of 'modern' inputs are often attracted to any new intervention which promises to provide them with such inputs. Mathieu Ouédraogo points out that wealthier farmers, who initially saw the presence of a white person as an indication of a source of inputs, subsequently withdrew when they realised that it was mostly about how to use a water tube to implement a traditional technique.

Furthermore, better-off farmers were not very keen on working in groups, because they were capable of carrying out household work using family labour or hiring the labour of poorer farmers. As a result, they were initially marginalised, because they saw no point in leaving their own farms and working for others on a rota basis. It was only when such farmers gradually saw the benefits that they became interested, and began to invest resources in improvements to their fields. For poor farmers who invariably lacked adequate labour, working alone had little to offer. They lacked the means to work alone, but they hoped that through group work they would be in a position some day to treat their own farms. They continued to work, although the project did not promise anything.

Be flexible

The early years also revealed that development agencies have perhaps something to learn from Chinese acupuncture: to treat an illness, you do not have to stick the needles where the pain is felt. PAF had started its research exercise as a way of contributing to the process of halting environmental degradation and restoring tree cover. Its agenda was to get farmers to plant local species of trees. However, the initial objective of agro-forestry had to be de-emphasised. It was realised very early on that farmers were more interested in exploiting the benefits of *diguette* construction to improve crop yields, when they saw the immediate impact on cereal-crop yields. As Halidou Compaoré of the village of Ninigui explains, 'It's not that we don't appreciate the value of trees or are necessarily opposed to planting trees. However, if I die from hunger, who is going to look after the trees I have planted?' A companion put the problem differently: 'We have two thorns: one in the foot and the other in the backside. Help us to remove the one in our backside first. Then we can sit down to remove the one under the foot ourselves.'

Reafforestation in fragile eco-systems and among communities struggling for survival faces several problems. Trees that are planted may have potential economic value, but they take several years to produce any dividends. Furthermore, in an area of rapidly declining pasture, planted trees are seriously threatened by browsing animals. The problem of protecting trees from animals, or preventing animals from eating up young trees, is one that the traditional system had still not found ways of resolving. There was the practice of putting animal dung under young plants, to dissuade animals from eating the leaves. However, as goats, sheep, and cows become desperate for pasture, they are no longer deterred by the prospect of eating their own droppings. Moreover, as the seedlings grow up, the deterrent smell wanes, and the animals feast on the young leaves.

Therefore, when the project was reoriented to meeting farmers' needs, it was in recognition of the fact that although reafforestation remained an important objective, it could not be achieved simply by asking farmers, struggling to survive, to plant economically beneficial trees. The initial goal of tree planting had to be re-examined, and ways found to integrate it into the very process by which households secured their household food requirements. In this process, farmers were interested in the technique mainly to harvest run-off water, control the expansion of degraded land, or (better still) reclaim the manure and organic matter being lost from their

Table 2.2: Impact of diguettes on cereal yields

Year	Control plot		Treated plot		Annual rainfall (mm)	Increase in yield (%)
	No.	Yield (kg/100m^2)	No.	Yield (kg/100m^2)		
1981	3	5.10	14	8.57	692	68
1982	45	4.42	47	4.95	421	12
1983	37	2.95	63	4.18	413	42
1984	72	1.53	74	2.92	383	91

(Source: Reij 1987)

fields, in order to improve productivity on their cereal farms. Table 2.2, produced after the project had measured the impact of *diguettes* between 1981 and 1984, shows that local farmers were justified in their optimism.

The impact of expatriate staff

The early phase of PAF also highlighted the problem of how to begin working within a village community, using non-local staff, especially expatriate workers. What impact would this have on the future development of a project? In the case of PAF, the initiation, development, and transfer of management from expatriate to local staff was an unusually smooth process. Mathieu Ouédraogo, a local member of staff, was involved at the initial stages of research and was subsequently trained to assume management responsibility. It was a promising start. Yet nearly all the villages where the project started no longer form part of its zone of intervention. No individuals from these areas are currently involved in the project's activities.

According to Mathieu, who is now the project co-ordinator, all attempts to mobilise the very villages which were once full of life under the preceding expatriate co-ordinator have failed. For example, in villages such as Kao, Koriga, Tugiya, and Bogoya there is no longer any project activity. These villages had to be abandoned for lack of support. There seems to be very little community spirit in these villages. The collapse of activities within communities whose active involvement gave the initial impetus to develop the project from a research exercise to an operational one is, to say the least, perplexing. After all, they were the first to testify to the effectiveness of *diguettes*. It was they who first appreciated the need to construct *diguettes* on their fields, and they should have been the backbone of the project.

It is not easy to recapture the exact circumstances of the early days. Nevertheless, we can speculate that the failure of the project to continue working with the first villages could be due to several factors. These relate to the nature of Village Groups or local institutions, the way in which local communities were mobilised, the role of foreign aid in an area with a long history of external assistance, and the involvement of expatriate staff.

The history of Village Groups can be traced to French colonial intervention. During the colonial era, they had little to do with the interests of the rural community. Independence did not seem to change their character in any significant way. In many cases they were no more than an association of individuals formed around a prominent village figure. The involvement of such a person ensured the participation of others, and his lack of interest often led to the collapse of the group. Therefore, when everything appeared to be going well, success was sometimes due mainly to the interest of one or two individuals. Since these institutions were promoted by government as channels for credit to the rural community, they did not develop a capacity for functioning beyond the goal of looking for credit. Although it is difficult to ascertain the exact nature of Village Groups within the project area at the time, by and large it is safe to conclude that most village-level activity which involved both external agency and government interventions lacked an independent momentum. Villagers often took part in such activities mainly to please those who had encouraged them to participate.

Another reason for the loss of interest among the early participants could be the impact of excessive foreign intervention in the Yatenga province. As a local researcher from the University of Ouagadougou puts it, 'The people of Yatenga know what to say and do when there is the potential for foreign assistance. Academic researchers and representatives of NGOs on programme-identification missions have criss-crossed the entire province and asked all the questions imaginable. Every villager knows what answer to give.'

Excessive foreign-agency intervention reinforced the loss of self-confidence which was the legacy of colonial conquest, and strengthened the view that solutions to problems within the community lie outside the local community. This is reinforced when an expatriate is perceived as the problem-solver. Despite the remarkable attempt by the initial PAF team to enhance the confidence of the local community in their ability to find home-grown solutions to local problems, the cessation of work in the initial villages suggests that it was probably not as successful as had initially been believed.

Mathieu Ouédraogo argues that the local people's initial impression, when his expatriate predecessor was around, was that they could get assistance, in the form of either food aid or inputs. As a result, many groups worked mainly to impress. When the expatriate co-ordinator left without their hopes being fulfilled, his replacement by a local worker (Ouédraogo himself) served only to dampen their spirits. A common refrain was, 'Ah! the son of so and so: he will let us down, for sure.' Villagers would be quick to point out that when expatriate staff were around they did get inputs, but now that their own brother was in charge, he no longer took care of them. Yet it was not as if there were actually many inputs on offer, since the emphasis of the project was on popularising the technique.

In looking back at the early years, it is important to acknowledge that making an unfavourable comparison between the performance of a new project manager and that of the previous one is a universal tendency. Similarly, there is a tendency for some new managers to attribute problems that probably emanate from their own inadequacies to lapses of their predecessors. Nevertheless, the involvement of expatriate staff does bring both benefits and dangers. Expatriate staff can often reinforce confidence in local ideas and facilitate the adoption of new techniques. Since they are often perceived as knowledgeable, their endorsement of a technique can encourage its adoption. They easily rise above inter-clan or inter-village rivalry to reach out to those most in need. They are less subject to local pressure to provide assistance along kinship lines. As such, they could be in a better position to direct resources towards areas of genuine need. Also, local project managers might be more prone to work with influential people within the rural setting. Under village conditions, project managers of an externally-funded

NGO are usually part of the rural elite. As such, they could easily find themselves responding more readily to the demands of the better-off than to those of the poor and more vulnerable sectors of society.

On the other hand, the involvement of expatriate staff can easily undermine the viability of local institutions. By being initiators of change, they reinforce the belief that change can come only from outside. Many expatriate staff do all they can to identify with the local communities within which they work. They very often have to do so through particular individuals, seeking out for preference the most enthusiastic people. But through close identification with such people and their households in the village community, they unwittingly create conditions which allow the transformation of these individuals into unofficial spokespeople of the village community. They become rallying points and act as channels for the transmission of aid into the community. They tend to claim credit for any progress that the village makes, and often see themselves as indispensable. Subsequent attempts to de-emphasise their role in favour of group activity or institution building often lead to disenchantment among groups that were previously organised around such individuals. Furthermore, there is the problem of households closely identified with an expatriate member of staff being perceived (often wrongly) by other households to have received more material assistance than the rest. This can sometimes create unwarranted resentment or lack of interest among other households.

It is in this context that PAF has expanded, extending its activities beyond the construction of *diguettes* to include other soil-improvement measures, and reintroducing reafforestation as an important item on its agenda. Beginning from 1985, full responsibility for the development of the project was assumed by local staff. With this transformation, Oxfam firmly joined the community of NGOs operating in the Yatenga province.

3

The scope of PAF today

Objectives and priorities

PAF seeks to distinguish itself within the NGO environment of Yatenga province mainly through its philosophy of intervention, its emphasis on modest or small-scale initiatives, and the priority which it gives to co-operating with State and parastatal agencies operating in the province. PAF claims that it is a project engaged in a process of constant learning and development, in conjunction with the Village Groups it works with. It says it does not regard itself as having any ready solutions to rural problems. It therefore operates with flexibility to allow practicable solutions to evolve out of the active participation of beneficiaries. In its view, it is this philosophy which has guided the project's development.

Its essential objective, it argues, is to strengthen grassroots institutional capability to understand and assume responsibility for finding solutions to local socio-economic development problems. In Yatenga, this involves helping local people to investigate the causes and consequences of environmental degradation, identifying the most sustainable way for natural resources to be exploited, and training farmers in new skills. It means supporting the growth of activities aimed at restoring and maintaining vegetation cover. To ensure that there is a collective learning experience and the sharing of learned skills, PAF aims to strengthen Village Groups, in order to enhance their ability to achieve the objectives they set themselves. This is done despite the fact that the project would ideally like to work with whole villages, so as to avoid some of the division which arises from inter-group rivalry. The achievement of these objectives, the project argues, is better realised when there is a spirit of co-operation between internal and external actors. Internally, the project promotes collective activity among and within groups.

The project has sometimes worked with the whole village as a group, when it is appropriate. For example, in the village of Longa, the people were mobilised partly due to the fact that the Chief of the Village, Youssouf Ouédraogo, took the lead in collective work. (At this point it should be stressed that the family name Ouédraogo is very common in Burkina Faso, and neither Youssouf or any farmer or official quoted in this book who bears this family name is related to Mathieu Ouédraogo, the current co-ordinator of PAF.) As a former employee of the failed GERES experience (described earlier), Youssouf was introduced quite early in life to the need to protect the natural environment. In the village of Goumba, it was an atmosphere of festivity and group solidarity which attracted many people to the activities of the project. As Mahama Guiro, a village representative, explains, 'We enjoyed ourselves so much while working that those who did not join felt that they were missing something. We even went further, and constructed *diguettes* on the fields of reluctant farmers. When they saw the result, they could not help but join the Village Group.' Members of Village Groups in the two villages mentioned here point out that they had been inspired mainly by the success achieved through *diguette* construction in the neighbouring village of Noogo. Mathieu Ouédraogo claims with pride, 'Today PAF no longer needs to

The scope of PAF today

run after villagers. Members of village communities themselves now seek us out.'

In promoting group activity, the project claims that its priority is to help households to improve their living conditions through cost-effective, profitable, economical, and manageable techniques. Food production and access to cash income, it argues, are at the heart of any attempt to ensure food security. Consequently, particular attention is given to measures which improve food productivity and increase access to incomes. Developing inexpensive techniques is based on the view that if development is to benefit the people, if new techniques are to be adopted by rural farmers, then the adoption of such techniques must not undermine the very basis of survival in rural households.

Governments and external agencies sometimes forget that an input is good not simply because it has proven to be capable of increasing food output. If local farmers have to sell off their assets to acquire such an input, then it inadvertently destroys the ability of households to cope with difficult conditions. Such an input is a danger to the very society that it aims to help. Weather conditions play an important role in determining rural productivity. Unfortunately, few inputs have been designed to suit all weather conditions. It is within this context that PAF can justify its professed goal of promoting techniques that can be afforded by farmers without undermining their ability to cope with dramatic changes in social conditions. Certainly, some inputs are indispensable but beyond the reach of some farmers. Many farmers cannot easily afford basic implements like carts, pick-axes, and ploughs. Nevertheless, the decision to introduce any input should always be informed by the degree of stress that any acquisition will impose on rural households. Where it is obvious that the acquisition of such an input will impose an undue strain on households, the project works out a scheme which will enable households to spread out their payments.

PAF argues that it makes no pretensions to being capable of bearing the whole burden of development. Staff say they

Table 3.1: Agencies with which PAF co-operates

Agency	Areas of Co-operation
Regional Centre for Agro-Pastoral Development (CRPA)	Agricultural Services Division (SPA): Programming/execution of tasks, follow-up/evaluation, education/training.
	Animal husbandry division (SPE): training of farmers (theory and practice), follow-up visits to demonstration herds, confined animals, trying out other fodder varieties, etc.
Provincial Office of the Ministry of Environment and Tourism (SPET)	Developing village nurseries, education/ training, programming/execution of tasks, follow-up/evaluation, trying out other local tree varieties.
North Yatenga Food Growing Project (PNVY)	Training extension agents and farmers, organising exchange visits, sharing information through workshops and seminars, co-ordinating activities.

are aware that certain requirements for genuine development are beyond the scope of any one agency, especially one that sees its role as essentially that of facilitation. While they are aware of the limitations of State agencies, they point out the undeniable fact that NGOs come and go, but there will always be government — even if it is unable or reluctant to promote sustainable development. Consequently, the project works closely with State institutions. As Mathieu Ouédraogo explains, 'We do not intervene directly. We usually work through State technical services. In that way, farmers will get used to seeing the veterinarian or the agronomist or forester

The scope of PAF today

Agency	Areas of Co-operation
French Association of Volunteers for Progress (AFVP)	Educating and training farmers, complementarity on water and soil conservation techniques, co-ordination.
The Struggle against Desertification in Burkina Faso (LUCODEB)	Field complementarity in the area of animal husbandry, monthly meetings, co-ordination.
Provincial Office for the Organisation of Rural Society	Adult literacy for members of village management committees, organising seminars or workshops.
Provincial Office of the Ministry of Planning and Cooperation	Sharing information, co-ordinating and organising zones of intervention, training, programming of activities.

as the person to contact for help in the resolution of their problems. After all, it is the responsibility of government to promote the country's development.' It is within this context that PAF co-operates with the relevant State agencies in the province.

Areas of activity

PAF currently undertakes a wide range of activities, covering seven broad areas with direct relevance to an improvement in the well-being of rural households:

- agriculture
- the environment
- livestock
- training, supervision, and public education
- supporting farmers with basic production implements
- operating a revolving grain stock to help Village Groups and individuals in need
- research and development.

Agriculture: improving soil productivity

PAF's activities in the agricultural sector involve a continuous popularisation of techniques to increase soil productivity such as *diguettes*. The project also trains farmers in how to produce compost in their respective households for use as fertiliser on their farms. In this regard, farmers are advised to combine *diguette* construction with the use of compost so produced. The recommended technique for using compost is the traditional practice of *zay*, modified to make it more effective. The practice of *zay* recommended by PAF involves digging pits 5 cm to 15 cm deep in a catchment area just before the rains. Farmers are advised to heap the earth that has been turned up on the lower side of the slope, and then to follow the traditional procedure of putting a handful of well-decomposed manure into the pit. The mixture of compost and the rain water that collects in the pit as a result of heaping sand on the lower side of the slope improves the nutrient content of the soil. Crops planted in such pits therefore obtain more water and soil nutrients than those outside the *zay*.

The environment: afforestation

Restoration of vegetation cover still remains an important aspect of the project's agenda. Consequently, Village Groups are encouraged to establish local tree nurseries to support a programme of afforestation. Seedlings of trees most sought after by local farmers are raised at small village nurseries, for use by individual farmers. Farmers are encouraged to cultivate economically beneficial local varieties capable of

The scope of PAF today

Watering neem and eucalyptus seedlings in a tree nursery run by the Ministry of the Environment in Ouahigouya. The PAF team distributes seedlings to pilot villages.

withstanding the sharp changes in climatic conditions. Since soil nutrients and water concentration along *diguettes* are invariably better, the project encourages farmers to plant the local grass species, *pitta* (*Andropogon gayanus*), along *diguettes*, because of its varied utility to the local community.

Livestock rearing: animal confinement

PAF's experience with afforestation has shown that the most important threat to young trees, apart from low soil moisture, is browsing by animals. As a result, individual farmers are encouraged to confine their animals. However, such confinement is not aimed merely at protecting young seedlings. It also forms part of the attempt to improve productivity of the livestock sector and facilitate links between agricultural and animal husbandry. Farmers are encouraged and helped to cultivate, harvest, and conserve fodder to feed animals in confinement throughout the year. The confinement of livestock offers various possibilities for increasing productivity in the pastoral and agricultural sectors. It makes it easier to implement animal vaccination schemes, an essential requirement for rebuilding the livestock levels which were so drastically reduced during the drought and famine of the 1970s. It also increases the production of animal manure, which is preferred by local farmers for use as fertiliser on their fields. Local farmers argue that even if chemical fertiliser was affordable, they would be reluctant to use it because, in their view, it dries up the soil.

Training

Following from the project's philosophy, PAF sees its role as that of developing and spreading new ideas, through training, on how to utilise new techniques, and adapt traditional practices to changing realities. Training, supervision, and public education are therefore seen as central to the success of its programmes. PAF undertakes mass-education schemes on issues such as the need to conserve the environment, and the measures for doing so.

The scope of PAF today

Mathieu Ouédraogo, PAF Co-ordinator, and a diguette planted with pitta (Andropogon gayanus)

The revolving grain stock scheme

The revolving grain stock scheme was conceived as part of the project's strategy for mobilising rural society and helping poorer households to overcome the problem of labour shortage. In doing so, PAF, just like Groupements Naam, based its initiative on a traditional social institution known as *Song-Song Taaba* ('helping each other'). Yet rural mobilisation for productive activity is often hampered by certain constraints. *Song-Song Taaba* involves an agreement among households to offer each other labour services on a rota basis. It does not involve any cash payments and is intended to overcome the shortage of household labour during the peak periods of the farming season. Groups of young men offer their labour on each other's farms, on the understanding that they will be provided with food after the activity. A failure by any household to provide food is considered a violation of an unwritten agreement. If this continues, it undermines the very existence of the group.

PAF observed that some households could not participate in the *Song-Song Taaba* system, because providing food to a large labour force would deplete their household food stock. So they were incapable of constructing *diguettes* — the more so because such households tend to have a smaller work force. PAF therefore institutionalised a system of a revolving grain stock, whereby poorer households could borrow grain to provide food to those working on their fields. This is done on the understanding that after the harvest, grain taken out on loan will be replaced, with a low rate of interest which enables the stock to meet any increasing demand. The aim of the revolving stock is not only to help poorer households obtain labour. PAF identified the traditional practice of *Song-Song Taaba* as one worth preserving. It encourages group cohesiveness, solidarity, and collective activity.

Input provision

Apart from finding ways of helping farmers to overcome their labour constraints, the project does provide some basic

The scope of PAF today

inputs such as wheelbarrows, carts, pick-axes, shovels, and watering cans. *Diguette* construction using rock *bunds* is a laborious exercise. Gathering and transporting rocks is tedious, even with basic inputs such as wheelbarrows and carts. It is virtually impossible on a significant scale without these inputs. In undertaking to support farmers with such inputs, PAF considers that it is not neglecting the crucial importance of conscientisation and mobilisation. However, it is impossible to mobilise people except through concrete activity. There can be little activity and productive work if farmers have to rely solely on their own labour power to construct *diguettes* or undertake other activities.

To help poorer households to acquire such inputs, the project implements a co-financing scheme, guaranteed by the Village Group, whereby an initial payment of 25 per cent of the cost of inputs is made by farmers, while the remaining 75 per cent is initially borne by the project. However, a framework for a scheduled repayment of this amount, affordable by farmers and guaranteed by Village Groups, is usually agreed upon before inputs are provided.

As mentioned earlier, inputs sold to farmers are essentially rudimentary and are inappropriate for certain problems

PAF's one truck, on loan to farmers for the transport of stones

which confront local farmers. Since stones are becoming increasingly rare in villages, PAF acquired an Isuzu truck in 1989 to help farmers to transport stones from distant villages.

Research and development

As we have seen, the project sees itself as engaged in a learning exercise in conjunction with local farmers. Its research and development initiatives demonstrate a preparedness to explore and try out new techniques in response to new problems that are constantly emerging. For example, although it has a tipper truck, it recognises that the problem of the declining availability of stones cannot be solved by simply searching out new supplies of stone and acquiring more trucks. PAF staff appreciate the need to develop effective substitutes for stones. Also, there are problems which arise as a result of the confinement of livestock, such as sheep and goats. The project is conscious of the fact that it has no immediate answers to some of these problems, and may not be able to solve them on its own. As a result, it is exploring solutions to problems together with its partners: State agencies, NGOs, and Village Groups.

Geographical zones of intervention

PAF has shifted from conducting extension work in the largest possible number of villages to a focus on pilot villages. The decision to narrow down its area of intervention was apparently based on the argument that it is better to have a few successful activities in a sample of villages than to spread the project's resources over a wide area. Successful sample villages, used for demonstration, can be more cost-effective than attempting to cover a wide area. Within the context of Yatenga's many problems and the project's limited resources, trying to cover a wide area is like a drop in the ocean. As a result, PAF has begun scaling down its zones of intervention. An earlier stage which involved working in 14 districts covering 64 villages (20 pilot villages and 44 test villages) has been further reduced to

Table 3.2: PAF's zones of intervention, 1992

Zone	District	Village
Ouahigouya	Oula, Namissiguima, Koumbri, Thiou	Recko, Boulounga, Goumba, Longa, Noogo, Ninigui, Simbissigui
Gourcy	Gourcy, Bassi, Tougo	Ranawa, Fourma, Sologom-Nooré, Sorgho,* Rawoundé*
Seguenega	Seguenega, Kalsaka	Mogom, Magrougou, Iria, Gonsin
Titao	Titao	Rimassa,* Tanghin-Baongo, Tougri-bouli,* Wanobé

* Recent addition

10 districts and 20 villages (see Figure 3 on page 3).

'Pilot villages' are those where the project concentrates its efforts and resources. 'Test villages' are those where either the project's activities are at an initial stage, or there is only a minimal intervention. The principle behind the division of zones of intervention into pilot and test areas is derived from the fact that, given the scale of environmental degradation and the problem of generalised impoverishment, pilot villages are regarded as the areas where the best demonstrative effects can be achieved. Moreover, considering the limited staffing levels and resource base of the project, PAF is incapable of meeting any increasing demands.

The project is staffed by a 13-member team, all of them Burkinabè: the co-ordinator and his deputy, the administrator/accountant, a secretary/typist, two drivers for the project's vehicles, a manual worker, and two watchmen. The main responsibility for carrying out extension work rests with five animators (one of whom is a woman), who shoulder most of the burden of co-ordinating, supervising, and implementing the project's activities in the field. They are at the front-line of contact with Village Groups. At a rough estimate, these animators cover villages with a total population of more than 160,000 people, a substantial population to cover.

In rural development, it is important not to promise what cannot be delivered, argues Mathieu Ouédraogo, co-ordinator of PAF. He has a point. Far too often, rural communities' initiatives have foundered as a result of unfulfilled promises. This is another reason why PAF has reduced the number of districts where it is actively involved. This scaling down is also a recognition of the increasing number of NGOs undertaking similar activities — a situation which has both advantages and dangers.

Comment: competition or co-operation?

The situation in Yatenga raises certain dilemmas about NGO intervention. Although co-operation among NGOs has been talked about for a very long time, there seems to be very little to show for it. The boldest attempt to control and co-ordinate NGO intervention in Burkina Faso was undertaken during the administration of President Sankara (1983-1987). He criticised the ridiculous situation whereby external agencies would arrive in the country unannounced, and rush off to some village, only to realise that there was already another agency in that village doing exactly what they came to do. Concern for this situation led to the creation of a national office (BSONG), under the Ministry of Planning and Co-operation, to oversee the work of all agencies in the country.

The director of the provincial office of the Ministry of

The scope of PAF today

Planning and Co-operation acknowledges that the presence of NGOs in Yatenga has brought some benefits to the local people. He is, however, disappointed with the lack of co-operation among them: 'Each NGO strives to show others what it does. Our Ministry wanted to take some initiatives to facilitate greater co-ordination. We felt that these agencies themselves had seen the need for greater co-ordination. However, because of rivalry and personality conflicts, such a framework for co-ordination is yet to materialise,' he observes. 'It is still possible to find two different NGOs arriving on the same day and time, at the same village, to work with the same Village Group on the same issue. It is only greater co-ordination which can prevent this situation. Every NGO is aware of it, but each one of them denies responsibility for the lack of dialogue and blames it on others.' To a large extent, the director's concerns are justified.

PAF is entirely staffed by Burkinabè workers, unlike the Agro-Ecology Project (PAE), which still has some expatriate staff. Yet the fact that an NGO has an all-Burkinabè team has not been shown to be an important factor in facilitating greater co-ordination among them. It is certainly the case that irrespective of staff composition, philosophy, and methodology, there is invariably an unspoken rivalry and competition among NGOs. Such competition is often transferred to the village level. For example, there is some rivalry between Groupements Naam and Village Groups. Furthermore, PAF does not co-operate with the 'Six S' movement as closely as it does with other agencies. Perhaps this is due to their different scales of intervention and their desire to maintain their own identity; or perhaps it is the consequence of personality differences. Yet it may have something to do with Lédéa Ouédraogo's rejection of Village Groups, especially as organised by State institutions. However, the problems do seem to extend beyond the nature of Village Groups. In the village of Ninigui, Village Groups who were working with PAF were mocked for their lack of inputs by members of Groupements Naam interviewed for this book. These Village Groups in turn criticised Groupements

Naam for being lazy. 'Their leaders are paid and they live like government workers. They existed long before us, but they have nothing to show for it,' some villagers in Ninigui alleged.

It is important to probe beyond charges and counter-charges and see whether there is any basis for the rivalry. The picture which emerges is one of competition, of a type which is not particularly healthy. Part of the explanation may be found in inter-village or inter-clan rivalries, which are not uncommon in many developing societies. However, the multiplicity of NGOs has not improved the situation. In some cases, communities are split along lines defined by the respective agencies intervening in the village.

There are several reasons why there has to be both co-operation and rivalry. From an objective standpoint, the presence of many NGOs in Yatenga is unavoidable. No single agency has the resources or the skills to meet the diversified needs of the province. Even if it had, it is debatable whether developing an NGO conglomerate with responsibility for the entire province or the sole licence to intervene in the region would be beneficial to the people. It is certainly the case that the bigger an agency becomes, the greater the tendency for it to become alienated from the people. The more bureaucratic procedures are required to run a large agency efficiently, the less quickly the needs of village people can be responded to. Furthermore, the existence of many NGOs should contribute to the development of each of them: they often learn from each other, even if the process is not overt. For example, all agencies involved in soil and water conservation in Yatenga now utilise the water tube to construct *diguettes*, although they do not always acknowledge its origins.

There is also the fact that development aid to Africa is itself not co-ordinated. Each donor agency has its own priorities, which it often determines without consulting other agencies working in the region. There are rivalries among mother agencies and even among governments. It is naive to think that such rivalries will disappear simply because the agencies find themselves on a common terrain. They are competing in a market place where they need brisk salesmanship and marketing skills. Selling one's product involves competition and co-operation. Whether consciously or unconsciously, PAF is also engaged in selling itself — as witnessed by the numerous international visitors (rivalled perhaps only by the 'Six S' movement) which it receives.

There are also subjective reasons which engender competition or facilitate co-operation. Interpersonal relations among staff of NGOs can either facilitate or hinder co-operation. There is also the perceived threat that the existence of a rival agency poses, even if there is no real threat to the project's funding. The need to protect one's corporate identity often engenders a situation where communications are less than honest. Opening up one's problems to another agency is frequently perceived as providing ammunition for one's own destruction.

There is also the fact that NGOs offer jobs to people who are responsible, through the extended-family system, for feeding many more people. The survival instinct and the legitimate need to remain in employment often hinder co-operation. Facilitating greater co-operation is sometimes perceived as working oneself out of a job. This need to protect one's job is not a problem peculiar to Yatenga. It is applicable to mother agencies and to official aid organisations alike. Staff who process grant applications are well aware that jobs are at stake — their own, as well as other people's. A reduction in the processing of grant applications can create a case for cutting down staff levels. This scenario does not suggest that co-ordination is impossible or that competition is necessarily undesirable. But it helps to explain why the rhetoric of development agencies is not always backed up by practice.

Dilemmas of multi-agency intervention

As we saw in the first chapter, the problems of Yatenga are numerous. It is not within the capabilities of any one agency, acting alone, to solve them all, or help the people of Yatenga to find solutions to them. To this extent the need for multi-agency involvement is evident. What is not clear, however, is how these agencies should relate to each other.

A review of the agencies operating in the province reveals all the signs of competition and co-operation. The only unique organisation appears to be the Desjardins credit organisation, which runs rural savings and credit schemes. PAE is involved in popularising *diguettes* using the water tube. It also assists farmers to construct compost pits, and is involved in village tree-nursery development. AFVP is involved in educating and training farmers in soil and water conservation techniques. The 'Six S' movement is involved in all the above and much more. PAF focuses on soil and water conservation, as well as soil improvement. Yet PAF co-operates more actively with PNVY and AFVP than with PAE or the 'Six S' movement.

Many of the NGOs operating in the province will hardly own up to any rivalry. They are quick to argue that failure to co-operate arises from their different geographical areas of intervention. They also add that for co-operation to be meaningful, it has to be concrete and involve activities in the field. Regular meetings in the provincial capital, involving officials of various projects, often produce few useful results. When co-operation does not involve the beneficiaries (the local farmers), then the benefits of such collaboration are limited. When the meetings aim to share information about each agency's activities, they can often become a competitive display of rival claims. Rarely do they discuss the errors or inadequacies of various projects. As a result, there is often very little to learn.

The Desjardins project, although not involved in soil and water conservation, justifies its existence by criticising other NGOs, arguing that some agencies have conditioned people to expect external aid. This, it argues, hinders people from taking responsibility for their own lives. The PAE distinguishes itself by arguing that, unlike some NGOs which distribute food aid, it encourages people to rely on their own resources, even in an emergency. The 'Six S' movement and its Groupements Naam pride themselves on being home-grown, and claim to be attacking the problems of the people from a holistic perspective, rather than in an isolated manner,

The scope of PAF today

as some agencies do. They argue that the problems of rural households are interrelated, and it is impossible to make any meaningful impact by tackling one sector in isolation. PAF likes to see itself as a small-scale, effective, and flexible project.

It has to be said that all the projects' claims sound convincing. They do seem to know what they are about and are unlikely to operate as GERES did in the early days. Yet it is unsafe to conclude that they are therefore making a meaningful impact on the life of local people: how is such an impact to be measured objectively? Certainly, when people have easy access to clean water, they should be healthier. Yet there are many other factors which affect the health of communities. Even when it is possible to determine how people's lives have improved, it is not always easy to attribute the achievement to one sole NGO. Despite these difficulties, the chapter which follows attempts to assess the impact that PAF has made, especially in *diguette* construction.

4

Laying the foundations for development

The impact of diguettes

Farmers in Yatenga are not fools. The *diguettes* now so widely in use across the province may have earned the name 'magic stones', but local farmers are fully aware that yearly crop output depends on a combination of factors such as rainfall (especially its distribution), the presence or absence of pests, the type of soil and its fertility, and the availability of labour within each household. Nevertheless, some of the farmers interviewed for this book emphatically and categorically endorsed the value of *diguettes*.

Anecdotal evidence from farmers

Mahama Guiro, a farmer from Goumba in PAF's zone of intervention, for example, has no doubts about their effectiveness in improving soil fertility: 'The land which was once bare is now full of life. The ditches and valleys have been filled up. The *zippela*, once arid and waste lands, have become cultivable, thanks to *diguettes*. They slow down the erosive energy of flowing water and stop soil nutrients from

Laying the foundations for development

escaping. When we spread manure on a field, it stays there.' ... 'It is *diguettes* that have saved the village of Recko where most of the soil is *zecca*', says Hamidou Ouédraogo of Recko. ... Kassoum Ouédraogo of Simbissigui points out that pastoralists once again bring their flocks to drink from the creeks in the village fields. In the process, their dung fertilises the soil. ... Kinda Mady of Simbissigui expresses his confidence in *diguettes* like this: 'On a farm treated with *diguettes*, no matter the nature of the season, you will always harvest something.' ... Hamidou Konfé from Ninigui even claims to have enjoyed a second sorghum harvest as a result of *diguettes*: 'On wet clayey soils, after the first harvest, the stumps grow and develop again to provide a second minor harvest.'

The results of the questionnaires administered also show that household heads, questioned about their views on the impact of *diguettes* on crop yield, had no doubts about their positive impact. Almost 100 per cent of all those questioned said they did observe improved yields in fields treated with *diguettes*. More importantly, there was no difference between pilot and test villages in the recognition of the impact of *diguettes*. This reflects the fact that *diguette* construction has been adopted by nearly all agencies working within the province. It shows too that the early trial results which led PAF to promote *diguette* construction were justified. In 1987, a survey of 15 villages in Yatenga showed that treated plots yielded 384 kg/ha, compared with 211 kg/ha on non-treated plots. A yield increase of 45 per cent for a mean annual rainfall of 438 mm was also observed for that year. In the village of Bidi in the north of Yatenga, closer to the desert, Lamachère and Serpentier, conducting a survey in 1988 on the impact of *diguettes* on millet, found an increase of 10 per cent on the higher sides of the slope and a 40 per cent increase downhill.

Farmers also point out that *diguettes* have made an impact on the agricultural cycle, which, they argue, is slowly returning to what they knew when they were young. They claim that because of improved water infiltration on treated

Sorghum growing on land treated with diguettes

plots, it is now possible to start sowing even with the first rains (no matter how sparse), since enough moisture is usually retained to allow the seeds to germinate. In other words, *diguettes* can alleviate the damaging effect of short intervals of drought on the farming system. 'It used to be necessary to weed hardened soil three times, to loosen up the soil for a meagre harvest. Today, it's enough to weed just twice, to get a good crop on the same soils. What's more, weeding is now much easier,' Mahama Ouédraogo of Longa commented.

It can, however, be argued that rural farmers, in their enthusiasm for *diguettes*, are probably exaggerating their overall influence on the farming system. Mahama Ouédraogo's views might mislead one into thinking that as a result of *diguettes* the workload involved in cereal cultivation has been reduced. But the number of times that crops are weeded depends a lot on the rate of weed growth, which is influenced by soil-moisture levels, and it also depends on the availability of labour. So while weeding may be much easier, because the soil is moister, *diguettes* may actually increase the number of times that weeding has to be done. Moreover, farmers' enthusiasm can easily create the impression that, irrespective of the type of soil and pattern of rainfall, yields would always be good if *diguettes* were constructed. There is, however, a multiplicity of factors which affect crop yields. Nevertheless, the claim that soil-moisture levels have improved is supported by the results of the field enquiry.

Diguettes and crop diversification

Increased soil-moisture levels seem to have facilitated crop diversification on fields, since most traditional crops (millet, sorghum, groundnuts, sesame, okra) now easily reach maturity. This therefore enables farmers to use some of their land to cultivate other crops. Table 4.1 shows that pilot villages indicate a higher percentage increase and a wider diversity of crops than do test villages. This could have a lot to do with soil-moisture levels, as well as the support given to farmers. Pilot villages attract more resources from PAF than do test villages. The combination of the impact of *diguettes* and increased attention probably explains the trend towards crop diversification. What is interesting is the

Table 4.1: Numbers of households growing particular crops on treated plots

Crop	Pilot villages		Test villages	
	No.	%	No.	%
Cereals	696	91	369	84
Oleaginous crops (oil seed)	135	18	27	6
Legumes	388	51	171	39
Tubers (potatoes)	18	2	9	2
Cotton	19	3	0	0

(Source: Results of questionnaires)

cultivation of cotton in pilot villages. Although it is still insignificant (only 3 per cent in pilot villages, but none in test villages), the re-emergence of cotton cultivation could be a pointer to increasing soil moisture and fertility. Cotton cultivation had virtually disappeared in Yatenga since the period when colonial administrators tried to encourage its cultivation. The main reason for its decline was the reduced availability of fertile land and water.

Of even greater interest is the high percentage of farmers in pilot villages (18 per cent, as against 6 per cent in test villages) who grow oleaginous crops and vegetables. Although cereal cultivation remains the priority, the increasing cultivation of such crops could suggest a potential for crop diversification, especially if they are also grown for sale. Women, in particular, point out the renewed potential that fields treated with *diguettes* offer them. Three women — Hawa Sanaa of Goumba, Bintou Tao of Magrougou, and Sarata Ouédraogo of Longa — all testify that they now grow okra (an important vegetable in the household food basket) around their homes. Previously, they could grow it only in low-lying areas away from the homesteads. As Hawa points out, 'The problem used to be that as soon as the okra matured, you arrived to pick some for your evening meal, only to realise that someone else had harvested your entire crop. Now our okra is within sight. What's more, the yields are better now than before.'

Diguettes and the natural regeneration of plants

Diguette construction was initially promoted as part of the attempt to restore vegetation cover in the region. Villagers recall how they used to be mobilised and supervised by forestry agents to grow trees, but to no avail, because of the lack of moisture. Now they observe a gradual restoration of vegetation cover. Many species, including *néré*, baobab, and sheanut trees, now regenerate by themselves. *Pitta* (*Andropogon gayanus*), which used to be rare, can now be found in many villages. Not only do farmers observe plant growth along *diguettes*, but they point to overall natural plant regeneration in their villages.

Laying the foundations for development

According to the results of the questionnaires administered for this study, 100 per cent and 99 per cent of respondents in pilot and test villages respectively agree that plant growth along *diguettes* is much higher. Furthermore, 99 per cent and 98 per cent of respondents in pilot and test villages respectively say they have observed some natural plant regeneration within their villages. Despite this overwhelming testimony, it cannot be concluded that afforestation should no longer be a priority, on the grounds that the vegetation of Yatenga is being restored naturally. The results do not indicate the scale and type of plant regeneration taking place. Nor can it be concluded that regenerating plants will survive the destructive consequences of animal browsing. Even if young trees do survive, it is important to bear in mind that trees take several years to grow. As such, the effect will take a long time to be felt. Furthermore, increasing population pressure on forest resources will continue to threaten any modest gains.

Eucalyptus trees, planted along a diguette in 1988, photographed in 1990

The scale of diguette construction

Farmers argue that *diguettes* have given them a source of hope for the future. 'The *diguettes* have proved their advantages to us; they themselves create awareness in us,' says Mahama Guiro. 'With *diguettes*, we renew our hope in agriculture and in the environment,' adds Ousmane Ouédraogo from Noogo. Souleymane Ouédraogo from Ranawa is well aware that *diguette* construction is a painful process, but comments: 'The suffering is worthwhile, because the burden that is going to feed you will not kill the one who carries it.'

Statements of this type inspire optimism. They show that the people of Yatenga are generally aware of the positive contribution of *diguettes*. What they do not show is the extent to which this general awareness is translated into the construction of *diguettes*. Table 4.2 assesses the rate of *diguette* construction in Yatenga.

Table 4.2: Number of households involved in diguette construction

Extent	Pilot villages		Test villages	
	No.	%	No.	%
All plots	133	17	44	10
Part	547	71	319	72
None	86	11	75	17

(Source: Results of questionnaires)

Table 4.2 shows that only 17 per cent and 10 per cent of respondents in pilot and test villages respectively constructed *diguettes* on all their fields; 71 per cent and 72 per cent of respondents in pilot and test villages respectively constructed *diguettes* on some of their fields; 11 per cent and 17 per cent in pilot and test villages respectively constructed no *diguettes* at all. Although a higher percentage of respondents in pilot villages compared with test villages construct *diguettes*, and a lower percentage in pilot villages do not construct *diguettes* at all, the difference is not significant. The results indicate that despite farmers' increased awareness, PAF's work has not affected the extent to which *diguettes* are actually constructed on fields. This is confirmed by the relatively equal rates of *diguette* construction in all villages. Nearly everywhere, the majority of farmers have constructed some *diguettes* on their fields. These results could also indicate that despite the influx of NGOs into the province, they have so far not succeeded in expanding the scale of *diguette* construction significantly. The limitation of these results, however, lies in the fact that they do not indicate, in the case of those who have *diguettes* on only some of their fields, the exact proportion of farmland which has had *diguettes* constructed across it. Also, some farmers do know about *diguettes* but choose not to build them, especially on or above *batanga* soils, because they are worried about the possible flooding of their fields.

Problems of diguette construction

The evidence suggests that there are problems encountered in the construction of *diguettes*. Stones are increasingly rare, and people have to travel longer distances in order to gather them. The scarcity of stones sometimes generates inter-village conflicts. Furthermore, since many people cannot afford to hire vehicles to transport stones from distant locations, and also often lack other inputs, many fields are often left untreated. Farmers often lack the tools needed for the construction of *diguettes*, such as wheelbarrows, pick-axes, and iron bars.

Table 4.3 shows that out of a sample of 1,209 households, 29 per cent owned bullocks, 23 per cent owned bullock ploughs, and 19 per cent owned at least a cart. However, only 7 per cent owned a wheelbarrow, an important tool for transporting stones needed in *diguette* construction. The limited access to the necessary inputs could be interpreted to

Table 4.3: Households according to type of agricultural equipment owned

Item	No.	%
Cart	224	19
Wheelbarrow	81	7
Spade	207	17
Pick-axe	207	17
Plough	272	23
Bullock	353	29
Others	92	8

(Source: Results of questionnaires)

mean that despite the in-flow of external financial assistance to the province, local farmers have not benefited significantly and therefore have little to show for the aid money pumped into Yatenga. It could also be interpreted to mean that since many NGOs focus on creating awareness and on training programmes, rather than providing inputs, and since farmers' income levels have not risen enough to enable them to purchase their own inputs, they are as a result unable to buy the inputs needed in *diguette* construction. But it has to be remembered that ownership of bullocks (29 per cent), for example, is just as much a status symbol and a source of wealth as it is a mere production input. Be that as it may, the lack of adequate inputs often retards the progress of *diguette* construction, and many farmers advance the lack of inputs as the main reason for the limited area that has been treated, naming transport difficulties as the most important factor.

Table 4.4 indicates that when respondents in pilot villages were asked to rank the problems confronting them in order of priority, 38 per cent felt that transporting stones was the most important problem facing them in *diguette* construction,

Table 4.4: Households' ranking of problems confronted in diguette construction

Problem	Pilot villages No.	Pilot villages %	Test villages No.	Test villages %
Transport	290	38	231	52
Lack of materials	92	12	71	16
Labour	84	11	29	7
Finance	58	8	24	5
Lack of time	66	9	15	3

(Source: Results of questionnaires)

while 12 per cent blamed the lack of materials (pick-axes, spades, etc.), and 11 per cent identified shortage of labour as the most pressing problem.

Within test villages, 52 per cent of respondents ranked lack of transport as their most pressing problem, compared with 38 per cent in pilot villages. The general conclusion to be drawn is that the lack of facilities for transporting stones is the most important problem facing the people of Yatenga. The higher percentage in test villages probably reflects the fact that they have not received much support from PAF, unlike pilot villages. The support for pilot villages, however, has not altered the nature of the problems confronting farmers.

... and some possible solutions

The results show that the problem of the increasing scarcity of stones has more to do with the distances which farmers have to travel to collect them (and therefore the need for transport equipment) than with the absence of stones in the province. None of the respondents mentioned the non-existence of stones as a problem. Nevertheless, Lédéa Ouédraogo,

President of 'Six S', thinks that lack of stones is a problem, but argues that the traditional practice of *zay* could be an important substitute, since it performs a similar function to that of *diguettes*. However, *zay* is most effective on *zecca* soil types. In the opinion of the provincial director of the Ministry of Agriculture, in areas not conducive to earth *diguettes* (for example in the sandy soils of north Yatenga), the local grass *pitta*, planted along the contours, can be a substitute. However, where soil conditions allow, earth *diguettes* should be constructed, to reduce the dependence on stones.

Local farmers seem divided in their opinions. Some farmers of Recko and Sologom-Nooré think that the soil will have recovered its fertility and tree cover will be restored before the problem of stones becomes really acute. Others think that earth *diguettes* constructed along the contours and supported by grass on the lower end of the slope could be a solution. Yet others suggest a two-layer planting of *pitta* along the contour. According to the chief of Longa, this could replace *diguettes*. These divided opinions suggest either that local farmers have not as yet been obliged to improvise and therefore come up with some solutions (even if inadequate) to the problem, or that no practical solutions as yet exist.

However, the actual or potential lack of stones due to transport difficulties is not the only problem associated with *diguette* construction. As in any construction project, regular maintenance is always necessary to ensure that *diguettes* serve the purpose for which they were built. This is even more necessary in the case of a rural area like Yatenga, where children often have to invent something to keep themselves amused. Yatenga children love to go hunting rats and hedgehogs in the dry season, when they have little else to do. *Diguettes* have become a fertile breeding ground for these animals, because they offer food and protection against predators. When children displace stones in their search for these animals, their preoccupation with the hunt usually makes them forget to replace them. In so far as stones are not replaced, they easily render *diguettes* ineffective. As a result, farmers have to make regular inspection visits during the dry

Laying the foundations for development

Repairing a stone diguette after a rainstorm

season, to ensure that the stones are in place.

Even when children do not displace the stones, *diguettes* still have to be maintained, particularly in the wet season. Clayey soils have a high water-retention capacity. As a result, water easily collects uphill, and plant life is killed off when there is a continuous and heavy downpour. This usually happens when *diguettes* fit too tightly (because of blockage over time) or are constructed too high. The owner therefore has to make regular visits to drain off water by removing a few stones from the *diguette* and replacing them later. It has to be remembered that *diguettes* are also meant to halt the process of soil erosion, not only to retain water, and as such have to be well maintained, even on soils with high water-retention capacity. Despite these problems, stone *diguettes* are by far the most resistant and durable system, and require comparatively less maintenance. However, they do require some accompanying measures such as planting grass along them if their full potential is to be exploited. Although *pitta* and thorny trees are being planted along *diguettes*, for example in the villages of Recko and Longa, the practice, though widely understood, is not common.

PAF's role in diguette construction

As we have seen, awareness of the positive impact of *diguettes* is generally widespread in Yatenga. Although it is difficult for PAF to claim any particular credit beyond its role as pioneer, Table 4.5 suggests that PAF's strength probably lies in the effectiveness of its extension methods.

Table 4.5: Number of households with knowledge of diguette-construction technique

Knowledge	Pilot villages No.	Pilot villages %	Test villages No.	Test villages %
Do not practise	37	5	78	18
Good knowledge	708	92	124	28
Some knowledge	22	3	180	41
No knowledge	3	0.4	60	14

(Source: Results of questionnaires)

The results show that 92 per cent of respondents within PAF's priority areas of intervention showed a good knowledge of the technique of *diguette* construction, compared with 28 per cent in test villages. Furthermore, only 3 per cent in pilot villages, as against 14 per cent in test villages, expressed ignorance of the technique of *diguette* construction. These results suggest that PAF's training has been effective. But the scale of the training is less important than its ability to convey PAF's message to farmers. This conclusion is justified by the fact that when respondents were asked whether they had received training, 60 per cent from pilot villages said yes, compared with 55 per cent in test villages — a rather insignificant difference.

Farmers confirm this conclusion when they describe with confidence the manipulation of the water tube and how to

Laying the foundations for development

build *diguettes*. Ousseni Ouédraogo of Ranawa finds the operation of the water tube and the technique of *diguette* construction so simple that it is the least of his worries. 'You need three people to operate the tube easily and correctly. First, unwind your water tube and leave it on the ground or hold it, if possible. Then from a bowl of water placed at a higher level, put one end of the tube into the bowl of water and draw in water with your mouth at the other end, making sure that no air bubbles are trapped in between. To avoid catching the air bubbles, incline the tube slightly as you draw in water. Then, place your two planks of wood side by side on a flat surface, making sure that they are straight and perpendicular. If you do not do this, you will not know whether there are air bubbles in the tube or not. If the water is not enough, draw in more water so that the levels reach the major marks on your poles. Once you are satisfied that all is correct, then you are ready to begin work.

'To determine your contours, you place one pole at one point (held by your partner), and you walk with the other

'The technique is simple — time-consuming, but worth it. It is our life.' (Ousseni Ouédraogo)

pole to a position on your field which your eye suggests to you could be level with the first pole. You read off your level and shout out your measurement to your partner with the other pole. If the water levels are not the same, you move it around a bit until you find a spot whose level is exactly equal to that of your partner. The third person then draws a line between the two poles using a *dabba* [hoe]. Once this is done, it is now your turn to remain steady with your plank, and your partner moves his to repeat the process. Slowly, you mark out your field, depending on how many *diguettes* you have decided to construct. If it does happen that two lines cross anywhere, then it means you have made a mistake somewhere. However, if you think the line is not perfectly straight, don't worry, it will not damage the *diguette*. After all, you can never draw a perfectly straight line on the ground. The technique is simple — time-consuming, but worth it. It is our life.'

Ousseni Ouédraogo's confident explanation bears testimony to the fact that if there is a problem with *diguettes*, it has little to do with not knowing how to use the water tube to determine contours. Nor is it to do with the actual building of the *diguette*, as Yacouba Ouédraogo of Sologom-Nooré explains: 'Once you have drawn your lines, then dig a ditch about this deep [pointing to his left palm: about 5 cm to 10 cm] with the *dabba*, or a pick-axe if the soil is hard. Heap the soil you have turned up on the upper end of the slope. Fix the stones that your wife has heaped into the ditch, making sure that they touch each other. Choose the stones which you think when fitted next to each other will be of equal height, about this high [using his right leg]. Ram in soil on the upper end of the slope, to avoid water seeping through the stones. If your field is small, then you should make sure your *diguette* is tightly constructed and has its mouths pointing towards the hilly side. When your field is large, you have to create a small space for a bit of water to flow over — otherwise, when the rains are heavy, you can create a small dam. You can do this by placing three lines of flat stone along higher sections of the *diguette*.'

Laying the foundations for development

'When your field is large, you have to create a small space for a bit of water to flow over — otherwise, when the rains are heavy, you can create a small dam.' (Yacouba Ouédraogo)

Yacouba completed his description of *diguette* construction by emphasising the need to reinforce the structures by 'planting grass or creeping plants along the *diguettes*. The grass will grow quickly, because the *diguette* has blocked rich soil beside it. This strengthens the *diguette* and eventually allows the stones to be removed if vegetation cover is restored.' So confident are farmers about their knowledge of these techniques that Abdoulaye Sankara of Gonsin reckons he has become a student who is capable of beating his teacher to the test. According to him, what he will miss most when PAF disappears is their truck, which helps him to transport stones.

Farmers' understanding of *diguette* construction is reflected in the type of *diguette* constructed. As Table 4.6 shows, in pilot villages, 92 per cent of household heads sampled said that they constructed stone *diguettes*, compared with 69 per cent in test villages. Moreover, only about 3 per cent of respondents in pilot villages constructed earth *diguettes*, compared with 8 per cent in test villages.

Table 4.6: Households according to type of diguette constructed

Type	Pilot villages No.	%	Test villages No.	%
Do not construct	37	5	93	21
Stones	705	92	305	69
Earth	22	3	37	8
Others	3	0.4	7	2

(Source: Results of questionnaires)

However, it cannot be concluded from this that construction of stone *diguettes* is mainly due to a higher level of training. Input provision could be a factor. Furthermore, despite the earlier conclusion highlighting the effectiveness of PAF's training methods, the project cannot claim all the credit. Although PAF is well known among villages sampled, it had been known mainly within a 3–5 year period.

In a sample of 767 households questioned in pilot villages, 90 per cent, as against 86 per cent of a sample of 442 in test villages, were aware of a project called PAF. Of those questioned in pilot villages, 56 per cent, as against 27 per cent in test villages, knew of the project within the last 3-5 year period.

Table 4.7 shows that 38 per cent of respondents in test villages had a longer (6–10 years) knowledge of PAF. This is probably due to the fact that test villages include the villages that PAF started with but has since abandoned. When further questioned about the source of their knowledge of PAF, 39 per cent in pilot villages, as against 37 per cent in test villages, attributed their knowledge to State extension workers, as shown in Table 4.8. In fact, more farmers were aware of PAF through the State agricultural extension services than through the project's staff. Furthermore, the

Laying the foundations for development

Table 4.7: Households according to length of knowledge of PAF

Number of years	Pilot villages No.	%	Test villages No.	%
1-2	35	5	92	21
3-5	427	56	112	25
6-10	207	27	169	38
11-15	5	1	9	2

(Source: Results of questionnaires)

majority of farmers attributed their knowledge of soil and water conservation techniques to staff of CRPA (the State agricultural agency) than to PAF.

Table 4.8: Households according to source of knowledge about PAF

Source	Pilot villages No.	%	Test villages No.	%
Village community	35	5	58	13
State extension workers	297	39	164	37
PAF staff	242	32	118	27
Trained peasants	7	1	21	5
Village Group	235	31	51	11
Radio	41	5	37	8
Others	8	1	5	1

(Source: Results of questionnaires. Note: responses are not mutually exclusive)

Table 4.9: Households according to source of knowledge about soil-conservation techniques

Source	Pilot villages No.	Pilot villages %	Test villages No.	Test villages %	Total numbers
No one	34	4	5	1	39
PAF	298	39	81	18	379
CRPA	413	54	333	75	746
Other NGO	22	3	23	5	45

(Source: Results of questionnaires)

In pilot villages, 54 per cent of respondents in pilot villages attributed their source of knowledge of soil conservation techniques to CRPA, as against 75 per cent in test villages. There was nevertheless a higher percentage (39 per cent) in pilot villages who acknowledged PAF as the source of their knowledge, as against 18 per cent in test villages.

If winning laurels were the aim, then most of the credit should actually go to the staff of CRPA. Perhaps, for once, a State agency is being credited for some good work done. Nevertheless, considering PAF's close collaboration with State extension services, especially the contribution it makes in terms of bearing part of the running costs of State extension workers, it surely deserves to take a significant share of the credit. After all, it is the support of PAF which has enabled CRPA to be more effective within the project's zone of operations. If, however, the goal is not to receive praise, then both PAF and CRPA should feel relieved by the high level of awareness among local farmers of techniques of *diguette* construction. More important, however, is the fact that an overwhelming majority of farmers rank soil and water conservation and natural regeneration as an issue of utmost concern to them. This awareness is also illustrated by the fact that in the village of Ninigui, each woman has planted a tree near her compound.

Table 4.10: Households according to prioritisation of soil conservation and plant regeneration

Prioritisation	Pilot villages No.	%	Test villages No.	%
Not a priority	25	3	0	0
Less priority	4	0.5	12	3
First priority	736	96	430	97

(Source: Results of questionnaires)

Yet the problems confronting the farmers of Yatenga go beyond the issue of harvesting water in order to improve the water and soil conditions. Much of the soil, unless enriched quickly with soil nutrients, is unlikely to recover its fertility, given the absence of fallowing and the problem of declining land holdings.

Beyond *diguettes*: PAF's complementary activities

Soil improvement

Despite the general knowledge of soil and water conservation techniques, the degraded nature of the soils plays a vital role in reducing yearly crop yields. It is for this reason that the project encourages farmers to produce compost in their households, to use in conjunction with their traditional practice of *zay*, as a way of improving soil fertility. Between 1987 and 1991, 375 farmers in 16 villages were trained in how to produce compost. Local farmers appreciate the qualitative difference between compost produced using the new technique and their traditional manure, as Mahama Ouédraogo of Magrougou explains: 'Before, we used to spread manure from our animal pen on our fields just before the farming season. However, that manure is dry, powdery, light and does not mix easily with the soil. It is easily carried off by water and wind. Its decomposition is slow and incomplete, which retards its effect. Compost, on the other hand, contains everything. It is well decomposed, fresh, thick, and sticks to the soil; it improves its structure, and raises its resistance to erosion. What's more, it replenishes the fertilising substances depleted through the annual crop harvest.'

However, nine out of ten farmers highlighted the lack of transport for carting compost from the household to the field as one of the most important problems. Moreover, lack of water and means of transport often means that women's work load is doubled, as they have to fetch water for household use as well as watering compost. It is not always easy to store up only waste water for use in watering the compost. Hamidou Ouédraogo of Recko also complained about the dangers that producing compost poses. Because pits have to be emptied manually, there is always the danger of being pricked by rusted objects (with the risk of getting tetanus infection) or

Loading compost in Noogo village. 'Compost is well decomposed, fresh, thick, and sticks to the soil; it improves its structure, and raises its resistance to erosion.' (Mahama Ouédraogo)

being bitten by scorpions, which tend to breed in compost pits. This directly raises the family's health bill.

Farmers interviewed also pointed out that compost rich in animal manure is best. However, because of a decline in livestock holdings, they do not have enough manure to feed the compost pits. As a result, they make do with any waste, some of which does not decompose easily. It is partly in response to this problem, partly to protect young seedlings, and partly to increase the productivity of the livestock sector that the project encourages households in pilot villages to confine their animals.

Animal confinement

The head of the regional animal husbandry services argues that it is still too early to assess the impact of animal confinement on the environment, since it has only just been introduced. However, its success will depend greatly on the resolution of several problems. There is the problem of inappropriate traditional implements for harvesting fodder, which leads to a lot of wastage and destruction. There is also the fact that fodder harvesting takes place during the farming season, thereby increasing the workload of farmers. Furthermore, because fodder is harvested during the rainy season, there is not always enough sunshine to dry it in the open sun. Fodder beaten by rain easily decomposes. There is also the problem of building suitable fodder-storage silos in the household, to ensure that enough fodder is kept and in good condition for animals throughout the dry season. Despite these problems, Hamidou Kagoné of Ninigui, responsible for one of the demonstration sites, has no doubts about the benefit of confinement. 'Animals so confined are healthy, and they develop and reproduce better than animals left to wander about,' he argues. Nevertheless, despite the significant progress being made in pilot villages compared with test villages, animal confinement is not as widespread as might be desired.

The results of the questionnaires administered showed that 47 per cent of respondents in pilot villages did confine their

'Confined animals are healthy, and they develop and reproduce better than animals left to wander about.'
(Hamidou Kagoné)

animals, compared with 25 per cent in test villages. This suggests that the joint work undertaken by PAF and CRPA has been partially successful. Despite this progress, the confinement of livestock is still the exception rather than the rule, since most farmers do not confine their animals. This indicates that there are some major obstacles which still have to be overcome if the livestock sector is to be made more productive and serve as a major source of income. There is the problem of the high cost of veterinary services and agro-industrial products such as salt-licks. Until livestock is protected from the threat of annihilation through disease, confinement could lead to a catastrophe for many households. Animal diseases, although easily treated when animals are confined, can spread faster and wreck havoc if veterinary doctors are not available, or the cost of drugs is beyond the means of farmers. If agro-industrial products are not available at affordable prices, the productivity of the sector could be fragile.

Laying the foundations for development

There is also the exchange value of livestock and livestock products to be considered. There is a declining market for live animals in Côte d'Ivoire (the traditional market), while other external markets such as Ghana remain unexplored. The lack of a market for beef from Yatenga and neighbouring provinces is a national problem, and until secure markets are found for livestock, increased holdings arising from improved management of animal herds could impose undue strains on environmental resources. The declining market in Côte d'Ivoire for Burkina Faso's livestock is due principally to the dumping of subsidised beef from the European Community on the Ivorian market. These subsidies have made European meat cheaper and beef from the cattle of Burkina Faso more expensive. Meanwhile, the Ghanaian market remains out of reach, partly because of frosty political relations between Ghana and Burkina Faso, and partly because of past problems in the Ghana-Burkina Faso cattle trade. Some cattle dealers from Burkina Faso still claim to be owed money by Kumasi-based Ghanaian cattle dealers, from transactions carried out in the early 1980s.

Training and extension work

The project's philosophy emphasises training and awareness creation among village communities at moderate cost. Training, supervision, and public education have been central to any successes the project has achieved. PAF staff are so well known in their zone of operations that children use them as role models in their games. This is most likely the result of their frequent visits to pilot villages to talk to farmers about the respective programmes of the project. Table 4.11 gives an indication of the number of farmers who have been trained by the project. The figures provided by the project show that very few farmers, if the total number of households in Yatenga (about 67,000) is taken into consideration, have been covered by the project. This can, however, be explained by the limited zone of PAF intervention.

Table 4.11: Number of farmers trained by PAF, and nature of training

Period	Theme	Number trained	Villages affected
1983–88	Water tube	4,542	406
1987–91	Fodder harvesting	375	16
1987–91	Compost pits	375	16
1987–91	Management committee	192	16
1987–91	Fodder preservation	52	16
1987–91	Urea compost pits	52	16
Total		5,588	

(Source: Rapport d'Activité, 1992, PAF)

Villagers interviewed were of the opinion that, on the whole, training was adequate. Their major problem was with getting inputs. They highlighted their inability to purchase inputs on their own, because the activities they were mostly engaged in were not income-generating. Although they were interested in *diguette* construction, the most common agricultural inputs that farmers owned were bullocks and ploughs, which are not useful for transporting stones. To that extent, farmers were justified in highlighting the need for inputs.

Promoting Village Groups

Despite its desire to work with whole communities, the project also emphasises group activity, especially through Village Groups. It was for this reason that a revolving grain stock was institutionalised, both to assist poorer households and to facilitate group solidarity. However, although the results of the field enquiry do show a higher proportion of *diguette* construction in pilot villages being undertaken through group work, the findings do not constitute a firm basis to conclude that there is enhanced group solidarity. After all, *diguette* construction by its nature is heavily dependent on collective activity.

About 93 per cent and 94 per cent of households in pilot and test villages questioned had participated in group construction of *diguettes*. Yet this is no indication that the project has been successful in strengthening group solidarity. Perhaps the only indication of the role that the project has played in promoting group activity can be obtained by examining in detail the patterns of labour utilisation in *diguette* construction. Table 4.12 shows that only 21 per cent in pilot villages and 23 per cent in test villages relied on group support to construct *diguettes*.

Table 4.12: Distribution of households according to manner of diguette construction

Manner	Pilot villages No.	Pilot villages %	Test villages No.	Test villages %
Do not construct (1)	45	6	83	19
Alone with family (2)	219	29	226	51
Collective work (3)	159	21	100	23
Combination of 2 and 3	344	45	33	7

(Source: Results of questionnaires)

The majority of farmers (45 per cent) in pilot villages relied on a combination of both family and group labour to construct *diguettes*. However, only 7 per cent in test villages relied on the same combination to construct *diguettes*. What is interesting is that fewer families (29 per cent) in pilot villages, compared with 51 per cent in test villages, depended mainly on family labour to construct *diguettes*. This suggests a significant role played by PAF in organising farmers.

Nevertheless, it says little about the ease or difficulty with which poorer households constructed *diguettes* using group labour.

Revolving grain stock scheme

The revolving grainstock scheme was instituted by PAF in order to enable poorer households to utilise group labour to construct *diguettes* on their fields. The scheme was also intended to strengthen group solidarity. The results of the questionnaires administered to farmers show that the majority of farmers never made use of the scheme.

Table 4.13: Households according to use of revolving grain stock

Frequency of use	Pilot villages No.	%	Test villages No.	%
Often	20	3	4	1
Sometimes	73	9	15	3
Never	660	86	402	91

(Source: Results of questionnaires)

In pilot villages, only 3 per cent of respondents often resorted to the revolving grain stock as a way of organising labour for constructing *diguettes* on their fields. In contrast, 86 per cent said they had never relied on the scheme. There are several possible explanations for the low utilisation of the scheme. It may be that because other organisations, such as the 'Six S' movement, did sometimes operate food-for-work programmes in order to speed up the rate of construction of *diguettes*, poorer farmers did not understand why they had to do something similar by incurring a grain debt. Also, it may well be that many households in pilot villages had no reason to rely on the scheme, probably because they could provide food when others came to work on their fields. It could also

Laying the foundations for development

be that poorer households for whom it was intended were so poor that they preferred taking a grain loan to feed themselves, rather than use it to feed those helping in the construction of *diguettes* on their farms. However, considering the scale of the problem of food insecurity in the province, it is unlikely that there would be no poor households in pilot villages in need of grain. If this were the case, there would have been no reason to institute the scheme in the first place. It is more probable that some households needed grain to feed themselves, rather than feed those coming to help. In such a case, they would prefer asking for a loan of cereal from a relation or a neighbour, rather than from the scheme.

Yet another interpretation could be that the Village Groups which tried to operate the scheme did not run it well. This could have been due to poor management and poor repayments of loans taken, with a consequent and gradual depletion of the grain stock. Non-payments or seriously delayed payment can quickly undermine the viability of such a scheme. However, the results of the questionnaires administered showed that among the small number of farmers who have had recourse to the scheme in pilot villages, a higher proportion (66 per cent), compared with 28 per cent in test villages, paid back without difficulty.

Table 4.14: Households according to capacity to repay grain loans

Capacity to repay	Pilot villages No.	%	Test villages No.	%
Without difficulty	73	66	5	28
With difficulty	30	27	11	61
Incapable	8	7	2	11

(Source: Results of questionnaires)

Table 4.14 shows that, of the small number who sometimes did use the revolving grain stock scheme, 27 per cent in pilot villages, compared with 61 per cent in test villages, had difficulty repaying. Also, 7 per cent in pilot villages and 11 per cent in test villages did not manage to repay the grain loan at all. However, a comparison between pilot and test villages is not useful in determining the operation of the revolving grain scheme, since the scheme was almost non-existent in test villages. Furthermore, even the results from pilot villages do not provide a useful indication of the operation of the scheme, considering the small proportion who made use of it. In any case, the indications are that the scheme is not serving its purpose.

Mathieu Ouédraogo argues that low rates of repayment are due to poor seasonal harvests, which usually hit the poorest households most. He argues that PAF is caught in a dilemma: it is opposed to giving out food aid, yet it finds it necessary to operate a revolving stock to help poorer households to invite a pool of labour to help them in their farm work. It is poorer households who are most affected by labour shortages during the peak farming season. They are also the ones most vulnerable to lower crop yields. They rarely harvest enough to repay the grain loan taken and retain enough grain to see them through to the next harvest. As a result, it has not operated properly, and that threatens the entire scheme.

As Mathieu Ouédraogo points out, 'Even in villages where it seemed to be working well, there is usually no repayment when the season is bad. Sometimes, you see in the books that there is supposed to be a certain quantity of grain, but there is no grain in stock to show for it.' However, he thinks that the problem is not only a question of poor harvests. It has much to do with the way in which households manage their grain stock. Sometimes households are unable to ration out their cereal properly, hoping that the next farming season will be better. He argues further that the situation goes deeper into the very priorities of rural society. For example, when there is a funeral or social activity after the harvest, households do not stint themselves — a point endorsed by Lizeta Porgo, a woman of 65: 'In the past, a bit of *dolo* [local beer] and *tô* [millet porridge] was enough, but now you need a whole bag of rice, lots of *dolo*, and lots of *tô*. In the old days, funerals were occasions which brought people together, since neighbours would give help and support. Today it seems that if you want a good funeral, you must spend a lot of money.'

While Mathieu does have a point, it is worth mentioning that poor management of household resources is not unique to Yatenga society (or, for that matter, to Africa). European expenditure patterns during the Christmas season cannot be explained by the logic of need. It is an established practice to give gifts during the season, and families endeavour to do so even when they can hardly afford it. To that extent, the celebration of funerals and baptisms in Yatenga is understandable: they have important social significance. Against this background, what is probably more valid is Lizeta Porgo's argument about the increasing cost of living. However, until local consciousness and traditions catch up with the type of market economy introduced by colonialism, household resources will be managed within the framework of the social institutions which the self-provisioning farming economy has shaped over a prolonged period of time.

Comment: how strong are the foundations?

When in 1979 Bill Hereford decided to try out his observations from the Negev desert, the primary concern was to find out how rain-water could be harvested and used productively within the arid or semi-arid regions of Burkina Faso. The initial objective of growing trees gave way to increasing food-crop cultivation, partly in response to farmers' priorities, but also in recognition of the fact that it was impossible for such a small project to attempt to confront all of Yatenga's problems simultaneously. There was therefore a need to set priorities. It was in this context that training and promoting the construction of *diguettes* topped the list of priorities. The hope was that if it led to significant increases in crop yields, as initial trials had shown, it could constitute a firm foundation for the development of the province. After all, the local economy was and still remains primarily a food-growing farming system. Other components of PAF's work, such as training farmers in how to produce compost for use as fertiliser on their fields, encouraging farmers to confine their animals, and training them in how to harvest and preserve fodder for confined animals, were all complementary to the main goal of increasing the productivity of the farming system.

Since the intervention of the project, both the testimonies of farmers and the results of the field enquiry and other studies confirm that *diguettes* do help to increase crop yields. *Diguettes* increase the overall soil-moisture content and thereby reduce the risk from declining rainfall. Their overall impact has been slowly to restore the farming cycle to what it was before the drought and famine of the 1970s. As a result, farmers testify that it is now possible to begin sowing after the first rains. The increasing levels of soil moisture have also provided farmers with an opportunity to cultivate non-cereal crops which had been dying out. Women in particular now find it easier to grow okra nearer home than used to be the case. *Diguettes* have also enabled some natural regeneration of the vegetation of the province.

So the positive contribution of *diguettes* is recognised by all farmers, and they are certainly anxious to reap the benefits. However, this enthusiasm is not matched by the level of *diguette* construction. Several problems confront farmers in their bid to take advantage of this soil and water conservation technique. The most acute is that of the declining availability of stones. It is not so much the *absence* of stones, but rather the increasing distances that farmers have to travel to find them, and lack of adequate transport to convey them from distant areas to their fields. Only 19 per cent of farmers interviewed owned carts, the most basic input which can be used to transport stones over long distances; only 7 per cent owned a wheelbarrow. It was therefore not surprising that most farmers ranked the lack of transport facilities as their most pressing problem. In villages where

Laying the foundations for development

Only 7 per cent of surveyed households in Yatenga own a wheelbarrow

PAF is not very active, 52 per cent of farmers questioned considered it their most pressing problem. In the project's pilot villages, 38 per cent considered it their most serious problem.

Yet there does not appear to be a solution to the increasing rarity of stones. Although the traditional technique, *zay*, is useful for harnessing rain-water for use directly on crops sown, and local people recognise its utility in increasing crop yields, it is unable to retain significant quantities of rain-water over the whole farm. This is necessary if overall soil-moisture levels are to increase as the water table rises. Another alternative to stone *bunds* could be *diguettes* made of earth, if reinforced with the local grass *pitta*. But earth *diguettes* need greater maintenance and management than stone ones if they are to be effective.

Is extension enough?

Against the background of the problems discussed above, PAF's greatest impact is arguably the effectiveness of its extension work, rather than its ability to find solutions to problems confronted by farmers in the process of *diguette* construction. Although there is no significant difference in the numbers of farmers trained in pilot villages (60 per cent) and in test villages (55 per cent), 92 per cent of farmers trained in pilot villages, compared with 28 per cent in test villages, said they had a good knowledge of how to construct *diguettes*. Also, more farmers in pilot villages constructed stone *diguettes* than earth *diguettes*. Considering the insignificant difference in the proportions of farmers trained in pilot and test villages, and in the light of the large number of farming households in the province (about 67,000), it would obviously be beneficial if the project extended its extension work. PAF has so far trained about 5,000 farmers, since its inception, in techniques of soil and water conservation and soil improvement.

Despite this recognition of the positive contribution being made by the project, it is also important to emphasise the role being played by the State agricultural extension agency, CRPA. The majority of farmers have become aware of Oxfam's project within the last five years. Their knowledge of PAF has been acquired mainly through CRPA. This is understandable, given the active collaboration between the project and CRPA. Such collaboration has paid off, as indicated by the level of farmers' knowledge about soil and water conservation techniques, and the overwhelming number of farmers who rank these techniques, as well as the restoration of vegetation cover, as issues of utmost importance to them.

However, increasing crop yields takes more than constructing *diguettes*. Enriching the soil is necessary if farmers are to stand a better chance of reaping a good harvest. Thus PAF's promotion of the manufacture of compost for use as fertiliser is a step in the right direction. Local farmers recognise this, and are willing to manufacture compost within their households for use on their farms. There are, however, important constraints. The most important ingredient, animal dung, is not easily available, partly because of declining livestock levels, and partly because of the dispersal of dung, since most livestock are not confined. There is also the perennial problem of lack of transport facilities to convey compost from the compounds to the fields. As was the case with *diguettes*, PAF has not helped farmers in any significant way to overcome this problem. It does have a truck, but wheelbarrows and donkey carts are more suited to transporting compost to local fields.

The project hopes to solve the problem of the declining availability of animal dung by encouraging households to confine their animals. The aim, however, goes beyond improving dung supply. It constitutes part of the project's goal of developing the livestock sector of the farming economy, as well as reducing the destructive impact of allowing animals to roam freely. PAF hopes that if animals are confined and fed properly, they will be healthier, easier to treat, and will reproduce better. Despite these good intentions, the reality is that most households do not confine their animals, except in the farming season when goats and sheep are pegged. Only 47 per cent of farmers in pilot villages and 25 per cent in test villages said that they confined their animals.

A number of problems undermine the ability of farmers to confine their livestock. There is the problem of feeding them in confinement, which is compounded by the use of traditional fodder-harvesting implements, which are inadequate and wasteful. There is the issue of competition for labour time between fodder harvesting and farming activity, since both have to take place during the farming season. There are also difficulties associated with drying harvested fodder. Fodder harvested risks being beaten by rain, which leads to rapid decomposition. This problem is linked to the lack of suitable facilities for storing fodder, which are necessary if enough fodder is to be retained for feeding confined animals throughout the dry season. Other problems affect the livestock sector, such as the high cost of veterinary services (drugs and agro-industrial products such as salt-licks), and the declining exchange value of livestock.

In the light of all the problems discussed above, the earlier conclusion that PAF's strength lies mainly in its training efforts is justified. The most important problem, however, remains unsolved: the inability of farmers to translate such training into concrete benefits which will improve their material wellbeing. Lack of inputs remains by far the most important problem facing farmers. Yet it is an area where PAF has made little impact. As a result, the development of farmers' fields to enable them to increase their crop yields (and consequently their incomes) so far remains an unattained goal. Meanwhile, the lack of inputs is further undermined by the generalised poverty within the farming community, as well as the limited access to cash incomes.

Some of the problems facing farmers could be overcome if they pooled their resources together. PAF therefore encourages farmers to operate within Village Groups. However, while data show a high level of group participation in *diguette* construction, there is little evidence to indicate that Village Group activity is enhanced. There is no significant difference between pilot and test villages in the numbers of farmers who have relied on group support to construct *diguettes*. There is, however, a basis for arguing that since households in pilot villages rely less on family labour to construct *diguettes* (29 per cent) than do households in test villages (51 per cent), perhaps Village Groups are active.

Although *diguette* construction in groups indicates greater communal solidarity, it is no basis for drawing firm conclusions that Village Groups are more active in pilot villages. Furthermore, it is difficult to discern which stratum of village society has benefited more from the project's initiatives operating under the umbrella of Village Groups. Influential village members, such as the better-off, local chiefs, and the so-called enlightened (those allegedly well-versed in modern ways, often returned migrants), tend to dominate these groups. While there is usually an atmosphere of free discussion, it is true that the dominant ones are able to rely on the group to help them develop their own farmland. The less well-off do participate in the activities of the group, although they rarely benefit directly. Quite often, their own farms are neglected.

The revolving grain stock scheme was established primarily to help poorer households to utilise group labour for the construction of *diguettes* on their fields. Yet 86 per cent of households in pilot villages said they had never made use of the scheme. This may be due, perhaps, to confusion created by the operation of food-for-work programmes by other NGOs within the province. As we have seen, farmers may be reluctant to borrow grain to feed a labour force involved in *diguette* construction. Instead, they may find it better to use the grain obtained as payment in the food-for-work programmes to feed others coming to help in *diguette* construction. It may well be that some farmers do not consider the construction of *diguettes* worth incurring a grain debt, when they themselves need to borrow grain to feed their households. In such a case, they may prefer to borrow from a relation or a neighbour. Perhaps the low participation in the revolving stock scheme may be due to the simple fact that it is not operating properly. This could lead to non-payment and therefore damage the viability of the scheme.

The project coordinator of PAF, Mathieu Ouédraogo, argues that the poor repayments are due to generally poor harvests. While this may be the case, it still does not invalidate the possibility that the scheme is poorly understood and poorly managed. This is not uncommon when the initiative to start a scheme does not originate from farmers themselves. Moreover, despite Mathieu's argument, about 66 per cent of those in pilot villages who made use of the scheme said they had no difficulty in making repayments, while only 7 per cent said that they could not repay the loan. Although Mathieu does have a point in that farmers' promised repayments often failed to materialise, the verdict that poor harvests and poor management of household grain stock are to blame is contestable. In view of the fact that only a small proportion of households relied on the scheme for grain to support *diguette* construction, and in the light of the fact that it is not exactly clear if it was mainly poor households who made use of it, poor repayments cannot be explained simply by bad harvests or poor management of household grain.

In view of the role of *diguettes* examined earlier, can one conclude that they have increased crop yields significantly to meet household grain requirements? The results of the enquiry showed that only 43 per cent of households in pilot villages reported that yields from farmland constructed with *diguettes* did cover household food needs. Yet this cannot be interpreted to mean that 43 per cent of those questioned no longer faced a threat of hunger — since food crops serve as a source of income as well as food. Rather, it will be safer to argue that the majority of households still suffer cereal deficits, although they may have constructed *diguettes* on their farms. Despite the renewed hope that *diguettes* have brought to local communities, the threat of famine still looms. A lack of rain prolonged over a number of years could lead to a repetition of the experience of the 1970s, when famine affected Yatenga province. This is not meant to undervalue the positive contribution of *diguettes*. However, since there are major problems which hinder the widespread construction of *diguettes*, their overall impact on crop yields will be limited. Furthermore, until other accompanying measures (manufacture of compost, confinement of animals) bear fruit, until solutions are found to the declining availability of stones and lack of transport and other inputs, the hopes generated by *diguettes* will remain unfulfilled. Until then, the people of Yatenga will continue to endure the pangs of hunger, and their lives will revolve around the seasonal struggle to secure their household food requirements. The hope that *diguettes* will contribute to the generation and accumulation of wealth for the present and the future will remain a dream, unless solutions are found to the interrelated problems of Yatenga.

5

Coping with hunger

Survival strategies

In terms of its potential for long-term cereal production, Yatenga is probably at more risk than any other province in Burkina Faso. Its population density is high and its soil fertility low. In 1991, 40 per cent of its population was categorised as vulnerable to famine (according to a seminar held that year on the co-ordination and management of information for food security in Burkina Faso). This high percentage, though in a rather bad year, reflected a history of virtually endemic annual cereal deficits. According to the Regional Centre for Agro-Pastoral Development (CRPA), in the 1982/83 farming year, the cereal deficit was estimated to be 62,722 tonnes. In the 1984/85 farming year, it was estimated to be 61,734 tonnes. When famine loomed on the horizon after the 1990/91 farming year, the deficit was 69,227 tonnes, about 50 per cent of the province's food needs. Despite rapid increases in population, food output has either remained stagnant or declined.

For many villagers, the threat that was posed to their very survival after the 1990/91 farming year brought back memories of the drought of the 1970s. In the village of Noogo, 15 km from Ouahigouya, farmers recall their experience of that year with anguish. 'We suffered a lot from the food shortage. No family had any reserves, because the previous harvest was meagre. There was grain for sale at OFNACER [the State cereal marketing agency], but people had no money to buy it. Those who had cattle sold them to buy grain. As for those who had only sheep and goats, they went borrowing. The fact is that the price of small animals was so low and that of millet so high that you could buy a 100 kg bag of millet sold for 9,000 FCFA at OFNACER and 12,000 FCFA at the local market by selling one animal. You risked selling all your animals without being able to buy enough grain to feed your family'.

For the villagers of Noogo, the problem was clearly one of lack of income, rather than lack of grain. A survey of women to find out what was their most pressing problem during that year showed that 86 per cent of respondents complained of lack of food. Despite the acute need for food, many of these women had no access to cash income and had to struggle to raise money if they were to provide a meal (for their children, at least). For many of them, borrowing was not an option. Nearly every one else was in crisis. The only options left were either to sell off an item or to try gold-panning in the area around Kalsaka.

Table 5.1: Women's experience of the 1990/91 hunger

Type	No.	%
Lack of water	257	21
Lack of food	1041	86
Other	32	3

(Source: Results of questionnaires)

Table 5.2: Households' coping mechanisms during 1990/91 hunger

Coping mechanism	No.	%
Sold item	443	37
Borrowed money	44	4
Gold panning	446	37
Market gardening	117	10
Others	356	29

(Source: Results of questionnaires)

Table 5.2 indicates that selling property or trying out freelance gold-panning were important coping mechanisms. Some youths from the village of Noogo consider the gold hunt a worthwhile venture: 'Some of us have been able to buy bicycles. Others have been able to roof their homes with aluminium sheets.' Listening to the young men can easily create the impression that, after all, the people of Yatenga do have ready access to gold and are therefore in a position to accumulate cash incomes. Yet the same youths are quick to point out that it is essentially a secondary activity and that farming remains their major preoccupation: 'Gold is a question of chance. You can dig for two days without finding anything, whereas with *diguettes* you know what to expéct. We rely on them to live today and tomorrow.' The same young men point out that whenever they are aware of the schedule of activities for the construction of *diguettes*, they invariably leave the gold sites to return home to help.

To a large extent, it can be argued that the youth of Noogo are painting an over-optimistic picture of the gold situation. For the majority of households the sale of livestock remains their most important coping strategy. Even then, most livestock are owned by men. For women, the gold sites are mainly a chance to offer themselves as casual labour to

Panning for gold near Kalsaka. 'Gold is a question of chance. You can dig for two days without finding anything, whereas with diguettes you know what to expect.' (Young man in Noogo village)

'permanent' gold hunters. Their tasks usually include fetching water for sale to gold diggers, and helping to dig or move sand out of the dug-out areas. What they receive as income for their labour is often small in real terms, and mainly supplementary to other coping mechanisms. As such, gold does not constitute, as far as women are concerned, a very important source of revenue for the household.

The limited access to cash incomes derives from villagers' heavy reliance on agriculture, as a source of both food and income. Table 5.3 shows that sales of grain, livestock, and some handicrafts constitute the main source of cash income. However, when everyone is struggling for food, there is little income left to purchase craft products. Grain sales are an important source of income, because in time of food shortage, cereal prices are high, while livestock prices fall dramatically. Although households usually sell off some livestock, they are often reluctant to part with most of their

Table 5.3: Distribution of households according to sources of income

Source	Major		Secondary		Occasional		Not a source	
	No.	%	No.	%	No.	%	No.	%
Cereal sale	399	33	93	8	96	8	621	51
Animal sale	392	32	508	42	102	8	207	17
Handicrafts	128	11	110	9	182	15	789	65
Market gardening	91	7	96	8	26	2	996	82
Remittances	88	7	116	10	216	18	789	65
Gold	64	5	57	5	143	12	945	78
Commerce	32	3	84	7	90	7	1003	83
Wage labour	17	1	48	4	58	5	1086	90
Cotton	1	0.1	8	0.6	5	0.4	1195	99

(Source: results of questionnaires)

animals, which they consider their most important investment. Instead, they sometimes reduce their consumption of available grain, to enable them to sell some grain and purchase other more urgently needed items. This explains why, although 51 per cent of households said they did not sell grain, grain sales still topped the list (33 per cent) as the most important source of cash income.

On the subject of handicrafts, Iddrissa Sawadogo recalls: 'Our parents were involved in a number of different occupations including agriculture, pastoralism, weaving, dyeing of clothes and leather, and making *guienda*, the instrument which women use to spin cotton. When I was young, dyeing and weaving were the most important

Behind the Lines of Stone

'When I was young, dyeing and weaving were the most important activity, after agriculture. ... Today, we can buy clothes made by les blancs which are already bright and colourful.' (Idrissa Sawadogo)

economic activity, apart from agriculture. Today, the craft of dyeing has disappeared. It has become obsolete, because there are quicker modern methods and in the market we can buy clothes made by *les blancs* which are already bright and colourful. The art of weaving is now mainly a symbolic activity. These days, our only other economic activity is market gardening, but it does not bring in enough money.'

The socio-economic impact of PAF

In the light of these findings, any assessment of the socio-economic impact of PAF leads probably to only one conclusion: despite the good intentions of the project, it has made little impact on the life of the people in the project areas. The old cycle of good years and lean years and the looming threat of famine remain, and so do the limitations imposed by the absence of alternative sources of income. For anyone familiar with the province before the intervention of PAF, a casual return visit a decade after it started work would not yield any surprises. Certainly, there are lines of stone on many fields, but the people still probably look as hungry and poor as ever.

Yet such a conclusion would be unfair to the project. In terms of rural development, the time frame of PAF's intervention is too short to expect any dramatic results. Furthermore, Oxfam never really set clear socio-economic yardsticks by which its intervention was to be measured. Also, as Marceau Rochette points out (in his book *Le Sahel en lutte contre la desertification: Leçons d'experience*), it is hardly possible to evaluate the direct effects of soil and water conservation techniques on the life of the local people. At best, what one can do is to assess the positive impact of such techniques on the consciousness of local people. This will involve assessing the levels of increasing consciousness among the target communities, the possibilities that these techniques open up, and the extent to which these possibilities are realisable (by individuals as well as groups). The assumption, in doing this, is that utilising soil and water conservation techniques should lead indirectly to improved living conditions. When viewed against this background, it is undeniable that PAF has had a positive socio-economic impact.

The contribution of *diguettes* to increases in crop yield has been established in Chapter 4, and the central role played by PAF in the development of *diguettes* was highlighted in Chapter 3. If the attainment of self-sufficiency in grain is central to a better life for the people of Yatenga, then the harvest results for the 1991/92 farming year do give PAF considerable grounds for satisfaction. The 1991/92 farming year was considered a good year for most farmers in the Yatenga province. Rainfall patterns were favourable, and farmers judged that *diguettes*, among other factors, did contribute to the bumper cereal harvest. For the first time in a long time, the official records (despite their unreliability) confirmed that the province need not always be perennially deficit in food supplies.

Table 5.4: Cereal balance (tonnes), 1991/92 farming season

Population	Cereal requirem'ts	Production	Losses	Net output	Balance
566,536	107,642	164,200	24,930	139,270	+31,628

(Source: L'Institut Nationale de la Statistique et de la Demographie (INSD))

The cereal balance for that year was a surplus of 31,628 tonnes, compared with the more usual average deficit of about 30,000 tonnes. An increase of more than 200 per cent in the grain harvest did certainly bring renewed hope to the people of the area. What this cereal surplus meant in real terms to the life of local farmers is difficult to assess, given the fact that it was achieved against the background of persistent cereal deficits. Furthermore, from the available figures it is difficult to determine inter-village or inter-district differences in crop yields. Also, a cereal surplus in the whole province does not tell us which households benefited, and how. Nor does it show how PAF's pilot villages fared. Yet it can be said with some certainty that when the cereal harvest is generally good, even poor households can feel the impact. They can at least feel more confident, compared with times of widespread poor yields, that they may benefit from the generosity of richer households when they are in need. To this extent, the 1991/92 harvest did contribute to improved living conditions. Increased availability of food does have an impact on cereal price levels. Its contribution to lowering cereal prices can enable poorer households to purchase grain from the market without selling most of their livestock, as happens during periods of grain shortage.

However, in recognising the positive contribution of a bumper crop harvest, it is important to see the contribution of *diguettes* in perspective. As the chief of Noogo pointed out, '*Diguettes* can only retain water if there is any. The stones themselves do not produce water. During the 1990/1991 farming season, we treated our plots with *diguettes* and compost, but it did not rain, so we harvested nothing. The *diguettes* and compost alone are not enough.' While appreciating the contribution of *diguettes*, it is also important to emphasise that improved living conditions involve more than having enough to eat, especially when increased food does not necessarily mean that households eat more nutritious diets. Furthermore, a better life entails having access to good drinking water, and proper health and educational facilities. Above all, it involves increasing the range of options available to farmers. An increase in local farmers' opportunities to gain extra cash income can contribute to the satisfaction of other basic needs.

The limits of PAF's role

Several important activities, like the provision of health care, though crucial, are outside PAF's scope of intervention. Similarly, access to clean water is central to good health — but, again, this is beyond the remit of the project. Educational facilities have an indirect impact on living conditions, but the project has not intervened in this area. It is unjustified to criticise PAF for failing to tackle these issues, without entering into the debate about the most appropriate strategy of intervention in an area of widespread poverty. Should an NGO tackle a prioritised list of problems, or should it adopt a wider focus of intervention, and develop a capacity to respond to the changing needs identified by the people themselves? Organisations such as the 'Six S' movement prefer a more integrated approach, while others, such as the PAE and PAF, focus on a narrower list of issues. As Daouda Ouédraogo of PAF explained, 'Our goal was a limited one: to ameliorate what was undoubtedly a critical situation in the province. Whole villages were virtually disappearing in a massive rural exodus, generated by the degradation of the environment and reduction in crop yields, which threatened to produce famine. Our priority was therefore to restore the soil in order to increase agricultural productivity and thereby guarantee food security.'

Despite such justification, the farming system plays a role far beyond simply providing households with food. Furthermore, one does not have to grow food in order to be food-secure. In an increasingly cash-based economy, helping farmers to earn money could have a more direct impact on living conditions. To this extent the project, in failing to diversify its focus to include activities which can bring in cash income directly to households, has reduced the socio-economic impact it could have had. The province of Yatenga has become famous for its vegetable production. Many of the vegetables which appear in big urban markets are cultivated within the province. Market gardening has become an important source of income for many households. It enables them to purchase clothes, pay school fees, buy medicines, pay for social events such as weddings and funerals, and improve their homes.

Table 5.5: Evolution of market gardening in Yatenga province: Production (tonnes), 1981–1991

Item	81–82	82–83	83–84	86–87	87–88	88–89	89–90	90–91
Tomatoes	200	112	172	4	69	357	701	145
Onions	445	155	409	317	680	1733	2542	1891
Potatoes	334	323	235	249	118	539	751	488
Cabbage	593	45	243	16	181	986	1887	775
Carrots	57	110	57	7	22	220	622	
Lettuce	60	10	3	0.25		192		38
Green beans	35		76		78	24	225	

(Sources: ORD Rapports d'activités 1985 and Porgo 1992)

Coping with hunger

Market gardening is an important source of income for many households in Yatenga; but it depends crucially on access to water supplies

The importance of market gardening in Yatenga is also illustrated by the fact that there is increasing interest in the development of techniques for preserving vegetables. In fact, vegetable growing has contributed to keeping young people at home during the dry season. As Table 5.5 indicates, vegetable output varies widely from year to year. However, during years of poor rainfall such as the 1989/90 farming season, vegetable output increased dramatically. The output of tomatoes increased dramatically from 69 tons in the 1987/88 to 357 tonnes and 701 tonnes in the 1988/89 and 1989/90 farming seasons respectively. Similarly, the output of onions increased from 1,733 tonnes in the 1988/89 farming season to 2,542 tonnes in the 1989/90 farming season. Under conditions of poor rainfall and poor grain harvests, an opportunity to earn some income, no matter how unreliable, is certainly a welcome one.

There are undoubtedly certain risks involved in vegetable gardening. The ability of growers to find markets for their produce will determine to a large extent the level of income which market gardening can contribute to the household economy. Although attempts are being made to improve the preservation of vegetables, the ability to transport products quickly to the market area will determine the volume of produce which a grower might sell. There are also problems of competition from processed vegetables imported from Côte d'Ivoire. Above all, however, is the problem of obtaining access to water (mainly from small dams) to grow vegetables, as well as the necessary inputs needed for effective vegetable growing. Many of these problems are outside the scope of any one project, although a project like PAF could help farmers try to find solutions to some of them.

PAF and migration

Within this context, it can be argued that PAF, by failing to promote alternative sources of income, has not made a major contribution to reducing migration out of the province. It is, however, worth asking whether the project should in the first place be interested in discouraging emigration. In view of the high population density in the province, will a lower population not reduce the pressure on the area's natural resources? In answering this question, the most important issue to examine is not simply the process of emigration, but the segments of society that are leaving. The exodus of the active labour force (mostly young men) has created a situation whereby the survival of whole communities is threatened by the absence of productive labour capable of producing adequate grain, in what is essentially a labour-intensive farming system. It is this which makes PAF's impact on migration an issue worth examining.

Harouna Ouédraogo of Ranawa thinks that to some extent the project has encouraged people to stay in the province. He argues that many farmers will return to Yatenga and stay there, especially if they are able to combine PAF's techniques with market gardening. Certainly, Harouna has a point. With the worsening economic situation in Côte d'Ivoire and the absence of job opportunities in urban centres, a combination of PAF's activities and market gardening will prove a useful incentive for young people to remain in the province. In the absence of this combination, the project has made little impact in keeping them at home. If there is any reduction in migration, it is perhaps the hope of finding gold and the declining fortunes of the Ivorian economy which are proving to be more powerful factors. To be able, however, to assess the true impact of the project, it is necessary to look at the ways in which PAF has affected those who bear the pangs of hunger most directly: the women.

PAF's impact on women

When Madame Yabré tried to investigate the socio-economic impact of PAF for this book, she was often confronted with a certain bewilderment. Women would look completely perplexed when asked whether the project had brought any improvement to their lives. To a woman of Yatenga, the notion of a better life appears absurd — more so when it concerns a project whose activities involve carrying stones. As a woman, the issues that concern her directly and daily are whether she has enough fuel-wood and water to prepare her evening meal with. She has to walk increasingly long distances to fetch them. Yet these are not the areas where PAF is actively involved. The project's activities actually increase the workload of women, yet bring them very little reward, as Table 5.6 shows.

Forty eight per cent of women interviewed were sure that *diguette* construction increased their workload, compared with 12 per cent who felt that it did lighten their work. Furthermore, only 28 per cent of women questioned were of the view that *diguette* construction provided them with a source of revenue, or facilitated better interaction among them. Considering the tendency of local women to accept their role in *diguette* construction as part of their contribution to the survival of the household, 48 per cent is quite significant. Their increased workload arises from the fact that women usually bear the responsibility for transporting

Table 5.6: Distribution of women according to benefit from or inconvenience of involvement in PAF activities

Benefits or inconvenience	No.	%
Increased workload	583	48
Reduced workload	150	12
Source of revenue	338	28
Greater interaction	335	28

(Source: Results of questionnaires)

stones, frequently in baskets carried on their heads, to the fields where *diguettes* are being constructed.

The problem of fuel-wood

Roambo Tampoko of Wanobé has no doubts in her mind: 'As for me, I want my husband to take a second wife. If there are two of us, I will not have to suffer like this.' Roambo is particularly resentful, because she is finding it increasingly difficult to find fuel-wood. Her supply soon runs out, but she has neither the money nor the time to travel very far in search of wood. Despite the government's efforts to encourage the use of improved wood stoves to conserve fuel, many women still rely greatly on their traditional cooking methods. Of the women questioned, only 42 per cent said they used improved wood stoves.

The women of Noogo express their fuel-wood problem in these terms: 'In the past, wood was available. We had trouble choosing which wood to cut. Today, the situation is different. We have to walk more than 10 km in search of wood. And what wood? Only twigs. Good-quality wood is rare. Often we have to make do with millet stalks, since at least they are always available. Even then, because of poor rainfall, millet does not grow well and their stalks look more like fodder. The problem of fuel-wood makes our work more difficult.

Coping with hunger

'In the past, we had trouble choosing which wood to cut. Today, the situation is different. We have to walk more than 10 kilometres in search of wood.' (Woman in Noogo village)

When the fuel-wood is not good, we are compelled to sit by the fire to keep it alight, instead of doing something else.'

Nevertheless, there is an increasing recognition of the utility of improved wood stoves, as Fati Sawadogo testifies: 'Ever since I started using improved wood stoves, this bundle of firewood allows me to prepare food for my family of 16 for four days, whereas previously it used to last only two days.' About 4,000 improved wood stoves are estimated to have been manufactured throughout the province, of which 60 per cent are actually in use. However, the Swiss-funded project which was producing the stoves has wound up, and women are reverting to their traditional methods of cooking. Unfortunately, local women have not been trained to continue manufacturing wood stoves themselves, using local materials.

Food availability

Despite all these problems, no particular attention has been paid to women by PAF. Whatever benefits the project has brought have accrued mainly to households, rather than to women in particular. Women, however, do agree that as a result of *diguette* construction, their food problems have improved. As Mahama Guiro of Goumba pointed out, he now rations out more grain to his wives. This view is also shared by farmers in the village of Boulounga. 'This year [1991/92], for instance, there is breakfast for everybody. Even if we do not take it, it is available. But during the difficult years, even the children had no breakfast. They had to make do with one meal a day.'

Women interviewed agreed that when the rains are good and the harvest plentiful, then food is abundant. However, this does not reduce their worries. While grain is provided by the men, the women have to provide for the sauce that goes with it. If they cannot use traditional vegetables, which often grow wild in the district, they must buy the ingredients for the sauce. However, the men often fail to provide the money for ingredients. Yet the men are always the first to complain: 'Our wives don't know how to cook. It's always the same: baobab leaves or okra.' ... 'But they forget that the quality of the sauce does not improve simply because there is now more grain. You always need money to buy ingredients,' the women of Noogo reply.

The menfolk, on the other hand, argue that 'We can't kill a chicken or a sheep just to eat. We can't afford that yet. We rear animals so that we can deal with social obligations, like funerals and dowries.' This is not disputed by the women, who add that they are not expecting their husbands to sell either livestock or millet simply to provide them with money for ingredients used in preparing sauce. They argue that if only they had ways of earning cash income, they could always manage.

Women's role in diguette construction

Despite the increased workload on women that has resulted from their involvement in PAF-promoted activities, men think that it is justified — as Iddrissa from Gounga explained: 'Before PAF, only men were involved in preparing the field for planting. The work was not systematic. When I noticed water running across my farm, I blocked its passage with some stones. The same goes for spreading manure, although my children sometimes helped me. With the intervention of PAF, the work is too difficult for men alone. We need the help of our women.' PAF workers explain the background of women's involvement as follows: 'In some villages the situation was so difficult that the men emigrated, leaving only the women and children. The women were left to face life alone. They were the key agents of production, and were interested and very active. So we worked with them, to help them to survive in the absence of their husbands.'

Women themselves recognise that the construction of *diguettes* is important, and that they must give a helping hand. 'In the past we did not take part in farming. Now that the food situation has degenerated and the solution lies in constructing *diguettes*, we join our husbands on the farm. The food problem concerns all of us,' explained Aïseta

'We can't do anything without the help of our women. It is the men who chip off the stones from the hillside, but it is the women and children who carry them in baskets to the farms.' (Kassoum Guiro)

Ouédraogo of Titao. This is supported by the results of the field enquiry. In pilot villages, 83 per cent of women questioned had taken part in *diguette* construction, compared with 76 per cent in test villages. This is a reflection of the fact that *diguette* construction is a labour-intensive exercise. Since most *diguettes* are built using both family labour and labour organised through Village Groups, women tend to play a crucial part in *diguette* construction, mainly as transporters of stone.

Men are fully aware of the effects of *diguette* construction on women. 'When we men return from the farm around 2 pm in the afternoon, we stretch ourselves under the shelters to rest, while the women continue to work: otherwise we will not eat in the evening. We know that our wives really suffer. That probably explains why our meals are usually not well prepared,' Kassoum Guiro of Boulounga explained. Yet they see little they can do to lighten the women's workload, as Kassoum added: 'All we can do is to encourage them with nice words. We know what to do, but we don't have the means, for example, to construct water sources or install a grinding mill. There are more than 20 traditional wells, but they all dry up by December. If we try to deepen them, they may cave in, because of the soft nature of the soil.'

Kassoum recognises that women's increased workload is mostly due to *diguette* construction. 'We cannot do without the help of our women. It is the men who chip off the stones from the hillside, but it is the women and children who carry them in baskets to the farms and help us to arrange them. Some women have even been trained to use the water tube and survey the field before it is worked upon. If the women are absent, no work can be done.'

The role of women in *diguette* construction is borne out by their replies to questions about the areas of project-related activity that women are most involved in. Of those questioned, 69 per cent identified *diguette* construction as their most common activity, nearly twice as many as were involved in afforestation; far fewer were engaged in the manufacture of compost, and fewer still in livestock rearing.

Table 5.7: Distribution of women according to type of PAF activity with which they are involved

Type of activity	No.	%
Diguettes	838	69
Afforestation	456	38
Composting	360	30
Livestock rearing	213	18

(Source: Results of questionnaires)

Women's income-generating activities

The effect of this increased workload has been a neglect of some income-generating activities. Yet many women are constantly exploring new ways of coping with their increased workload and limited sources of income. 'We don't work in the fields every day. On field-work days, we quickly prepare food for the family very early in the morning and then join the men in the field,' Fatimata of Gourcy explained. Women in polygamous marriages rotate the task of preparing food, so that those with responsibility for preparing food are exempt from farm work on their day of household duty.

In the village of Ranawa, in a predominantly Muslim area, these arrangements are significant, given the fact that traditionally women are not expected to leave the home. Women in the village of Tougri formed a group to hire out their labour services to raise income to meet some of their social obligations, such as funeral ceremonies. Some women from the village of Noogo teamed up to establish a tree nursery for the sale of seedlings (running the risk that some of their husbands might purchase the seedlings and fail to pay up). Yet such activities do not as yet constitute major sources of income for women, as Table 5.8 shows.

Table 5.8: Distribution of women according to income-generating activities

Activity	No.	%
Commerce	174	14
Handicrafts	717	59
Bee keeping	0	0
Market gardening	133	11
Gold panning	327	27
Others	102	8

(Source: Results of questionnaires)

The most important income-generating activity for women was handicraft production. Fifty nine per cent of women interviewed identified this as their most important source of income, compared with 27 per cent who derived their income from gold sites. Only 14 per cent and 11 per cent relied on petty commerce and market gardening respectively for income.

In general, women are increasingly relying on agriculture as a source of income, as reflected by the large numbers of women who own their own farms on a permanent basis. And in this they are sometimes supported by their husbands. When women decide to cultivate their own farms, it is usually to generate some independent income. This is necessary because of their husbands' increasing inability to cater for other household needs beyond the supply of food grain. The fact that women cultivate their own independent farms does not exempt them from farming on collective household farms. Instead, it simply means an increased workload for women.

What is significant too is that 52 per cent of the women sampled had permanent farm holdings, and 90 per cent of those who clearly had access to farm land cultivated it on a permanent basis, receiving it on loan from their husbands. The extent

Coping with hunger

Table 5.9: Nature and source of women's access to farms

Source	Permanent		Temporary	
	No.	%	No.	%
Husband	564	90	65	10
Land priest	42	57	8	11

(Source: Results of questionnaires)

to which women cultivate their own farms on a permanent basis is an indication of the determination of Yatenga women to obtain independent incomes. The preparedness of their husbands to loan out family land indicates the men's recognition of women's need to generate extra income. In times of hardship, men no longer feel obliged to provide for their wives.

Despite women's active involvement in farm work, they do not own the productive resources needed in the construction of *diguettes*. Furthermore, less than half of the sample had access to resources traditionally identified with women's needs. Table 5.10 shows that while 46 per cent and 42 per cent had access to a grinding mill and a bore-hole respectively, only 18 per cent had access to a cart, an important input used in transporting loads.

Table 5.10: Distribution of women according to use of resources

Resource	No.	%
Grinding mill	555	46
Bore-hole	510	42
Cart	216	18
Rickshaw	96	8

(Source: Results of questionnaires)

The need to target women

Despite the centrality of women in the household economy, it is only recently that PAF has given any attention to women's needs in particular. Perhaps this has a lot to do with the nature of the project's activities. Perhaps it is determined by gender relations within the social system. Irrespective of the reasons for the project's delayed targeting of women, the notion that women can be easily targeted in the rural economy is easier said than translated into action. The most important economic activity revolves around cereal cultivation. This exercise has become a joint endeavour within the household. Certainly, male household heads control the allocation of cereal grain. However, gone are the days when they could use the grain stock as they wished, without endangering the survival of the family.

This does not mean that there is therefore no need to target women in particular. Most male heads of households have greater access to income than women, although women spend more of their income on household needs. In 1990, Bernadette Hannequin compared current household expenses in the village of Rana in the Mossi plateau. Her study showed that 33 per cent of male incomes is spent on food and items related to household subsistence, compared with 86 per cent of female incomes. Also, the poorest male reserved about 1,600 FCFA (about £3.00) a month for personal expenses, while the richest woman had access to only 1,320 FCFA (about £2.40) a month. Men's incomes derive mainly from their control of livestock, while the income of women is often earned from the sale of handicrafts, agricultural produce such as groundnuts, or the processing of sheanut fruit into butter.

Therefore, targeting women will have to focus on areas of activity which are completely outside the domain of traditional male control. The line of least resistance is usually to provide assistance which can reduce the workload of women, such as a grinding mill or access to water. PAF has, however, done little for women in this area. For a project like PAF which prides itself on avoiding the reinforcement of an aid-dependent mentality, helping women to acquire grinding mills could run counter to the project's stated objectives. Owning a grinding mill is the dream of every women's group. Yet grinding mills have a tendency to increase dependency, especially when not managed properly. Nearly all the inputs — fuel, spare parts, sharpening of the grinding stone — required for the efficient operation of the mill have to be imported, at a cost which cannot simply be added on to the unit cost for milling. Operating a mill on the basis of profitability could alienate many women from using its services. As a result, some subsidy will be necessary. Yet to continue to subsidise its operation means that loans for the acquisition of mills will take a long time to be repaid.

The provision of clean drinking water also requires an investment which is beyond the financial capability of many NGOs. Certainly, technical advice on how to dig improved traditional wells could be a solution. However, until the water table rises such that water will be available all year round, traditional wells do not necessarily offer a solution. Where the water table is low, it means that deep drilling is necessary if local communities are to be provided with water. Yet the acquisition and maintenance of the required drilling equipment involves costs which are often beyond the financial capabilities of the government of Burkina Faso.

This is not intended to suggest that the targeting of women should not be a priority. On the contrary, given the increased workload that women have to bear, there is a good case for developing programmes which benefit women. Income-generating activities such as market gardening or handicraft production are usually mentioned as useful areas of intervention. However, what tends to be forgotten is the location of these activities in the broader market economy. Unless markets for women's products are secured, such activities probably have only a short-term utility. The greatest dilemma which the project faces is whether, in seeking to address the needs of the people, it should incorporate a broader spectrum of activities. As soon as that becomes its focus, then PAF is on the road to becoming an NGO conglomerate, with the risk of overlapping with other agencies, and all the rivalry and competition for resources which that role would necessarily entail.

Comment: a targeted, sustainable enterprise?

When, in 1982, Oxfam took the decision to become fully operational in the Yatenga province, it was aware of the scale of the problems confronting the area. The problems were environmental as well as human. By choosing to promote *diguettes*, Oxfam hoped to make a direct impact on the life of the people. After all, food is the foundation of life. Nearly a decade or so after, it is time to ask what exactly PAF has meant for the people within the pilot villages. As we have seen, it is difficult to measure the socio-economic impact of soil and water conservation techniques, more so as the project never set itself clear indicators by which its progress could be measured. Despite this limitation, it is worth asking if the intervention of PAF has had a positive impact on the life of the people, and whether such an impact would be sustainable if the project withdrew from the province.

If statistics are to be believed, the experience of the past decade does not indicate that, in the area of self-sufficiency in food, any progress made by PAF will be long-lasting. In 1991, the province was categorised as the worst off in terms of long-term cereal production, with 40 per cent of the population deemed vulnerable to famine. The cycle of lean years interspersed with the odd bumper harvest continues to characterise the farming system. Furthermore, the people of Yatenga still find it difficult to earn cash incomes, and remittances from family members in Côte d'Ivoire are no longer guaranteed. The fevered search for gold in and around Yatenga has not made the province's food supplies any more secure.

Yet local farmers express their faith in PAF, and do co-operate with it in carrying out various activities. When they do so, they can point to the cereal surplus of 31,628 tonnes which the province attained during the 1991/92 farming season. While they recognise that rainfall patterns played a crucial role, they are also convinced that *diguettes* and the application of soil-improvement techniques played a part. Such a contradictory picture makes it difficult to draw reliable conclusions from the first decade of PAF's experience.

The easiest conclusion to be drawn is that a decade, in rural development, given the scale of the problems, is not really long enough for an assessment of the impact of a project. Moreover, PAF's intervention is on a modest scale, compared with those of other organisations. It would be more appropriate to take a critical look at the project after another five years before drawing any firm conclusions. Yet such a view ignores the fact that the impact a project has made or will make does not depend simply on time — although time does provide an appropriate framework to identify the strength and weakness of any activity. Whether an object will float or sink when placed in water is not determined simply by the time since it was immersed.

Targeting the poorest?

It is, however, safe, in any assessment of PAF, to say that there is one aspect that the project can feel proud of. In technical terms, its original objective of promoting *diguettes* has been a success. In social terms, however, it has not been successful. It has apparently failed to modify the social processes which impoverish certain sectors of society. On the contrary, PAF seems to have enhanced the ability of richer households to become more productive.

Households with a capacity to transport stones, those capable of purchasing the necessary inputs to improve their productivity, and those with significant livestock have found the techniques being promoted by PAF to be beneficial. Despite PAF's awareness of the obstacles which face poorer households in the adoption of techniques which can improve their agricultural output, it has not examined the problem in a serious way. The project has approached the issue of unequal access to productive inputs in a manner which suggests that it has not paid sufficient attention to finding ways by which it can better target its intervention. Although it has sought to promote Village Groups, PAF often prefers to work with whole villages, in order to avoid the rivalries that often do arise between rival village groups. In taking this approach, it sees the village community as homogenous, with common interests.

The reality is, however, different. This is most evident when we examine the project's relationship with women in its pilot zones. These women argue that, although they appreciate the role of *diguettes*, the project has made life more difficult for them. It has increased their workload without helping them to solve their problems. If the material well-being of children is a good yardstick for measuring the real impact of a project on a community, then we should pay attention to women who report that it has certainly not been any easier to deal with the cries of hungry children. Yet one can argue that in any development activity there is usually an unequal sharing of the burden. After all, the workload of women would not have been any lighter if they were not constructing *diguettes*. However, a burden is easier to bear when there are indications that it will be lighter in the future.

Ensuring sustainability?

There is another aspect of people's lives which the project cannot claim to have altered very much. There is no evidence to suggest that PAF is strengthening local capacities to cope with the hostile environment of Yatenga. The water tube did revolutionise methods of soil and water conservation. Yet it was not the creativity of local farmers themselves which led to its development: it was the preparedness of the project's pioneers to try out an instrument which was still at an experimental stage. The view that solutions necessarily originate from outside local communities themselves has barely changed. Nor does it seem that a foundation is being laid for it to change. Mahama of Goumba certainly thinks that the project has come to teach them, and that its mission has not yet been achieved: 'It is like when you teach a child to walk. You stretch out your hand to enable it to take its first steps. If you leave it at this moment, it will fall back to its sitting position. You have to guide it a bit before you leave it. We are like the children of PAF. PAF is teaching us to walk, but we cannot as yet walk on our own. It will come, but we don't know when.'

The most worrying aspect is not simply the fact that PAF is leading the villagers, but the fact that farmers like Mahama cannot say *when* they will be in a position to walk on their own. If his analogy is anything to go by, it is not self-evident that a child learning how to walk will eventually walk properly. If walking means the community learning to develop and improve upon its methods of coping with local problems, then the project, after its first decade, has not left behind or developed from within the society any unique skills in community organisation or mobilisation. There are grounds for fearing that after the withdrawal of PAF, the obstacles which currently hamper local farmers may become stumbling blocks.

6

Past, present, and future

In many ways, the decision of Oxfam (UK and Ireland) to take a critical and retrospective look at one of its most internationally acclaimed projects was a positive one — positive in the sense that it has provided an opportunity for Oxfam to examine its role through PAF and for PAF to learn the relevant lessons of its intervention in Yatenga. Oxfam can be credited with taking a bold initiative, and the hope is that any lessons learned will guide the project as it charts its future course.

As we have seen in the preceding chapters, in general terms PAF has a lot to be proud of — not that it can claim any particular unique achievement, but more importantly because (although its direct impact on household food-security is difficult to quantify), there is no doubt that *diguettes* have made a generally positive contribution to the farming system of the Yatenga Province. Yet what is important for the project is not so much the recognition of PAF's achievement, but a critical examination of its methods. It is here that the lessons for the future can be learned.

PAF has always prided itself on using a participatory approach to produce cost-effective and simple techniques which farmers can use. It has always emphasised its focus on awareness raising and social organisation, and its determination to meet the needs of the disadvantaged. These priorities, the project argues, lie behind its three-pronged approach to rural development, involving an integration of afforestation, agriculture, and animal husbandry. In terms of participatory methodology, the development of the water tube and the movement from a single focus on afforestation to a combination of cereal cultivation, animal husbandry, and afforestation are seen as an illustration of its commitment to this approach. So also is the shift from encouraging the construction of *diguettes* only on group farms to a concern for both individual and group farms.

PAF and participatory methodology

There is no doubt that the project has travelled this route and made the appropriate changes when they were deemed necessary. What is still not clear is exactly what PAF means by a participatory methodology. Is a project's ability to modify its activities according to farmers' interest sufficient proof of a participatory methodology? Is having a good rapport with the local community in which a project works a major indicator of a participatory approach? While much has been written about participatory methodology, it is important not to ignore the difficulties inherent in making generalisations across regions and between different epochs. Moreover, it is not evident that participation necessarily solves all problems. Take, for example, the case in northern Ghana where 60 co-operative members (men and women) voted for the village chief to be President, and agreed that he could take the largest share of collectively owned grain. The group members found nothing wrong in this reinforcement of inequality. Furthermore, there are situations where a non-participatory approach may be more useful, as the experience of an Oxfam field worker in Burkina Faso showed. The field worker played a leading role in exposing a village pastor who was diverting funds from a collective sheep-rearing project into his private bank account. Other members of the group were either unable or unwilling to take decisive action against the pastor. Being the 'light' of the community, and having assumed responsibility as the village negotiator with external agencies, he was perceived locally as representing the community's interests. Consequently, some members of the community were prepared to forgive him.

In citing these examples, the purpose is not to deny the importance and relevance of participatory methodologies. On the contrary, the aim is to emphasise the need for a more rigorous understanding and development of structures that ensure genuine and effective participation. To unravel issues of participation, it is useful to find out who exactly are participating. Has it been both men and women, young and old, who have been actively involved in the project's activities? How have they been participating? Was it through the continuous involvement of clearly defined interest groups in the project's work, or has it been mainly an episodic ritual of mass meetings and consultations involving Village Groups and local communities? If local groups have been participating, what exactly is it that they have been engaged in? Have they been involved in decision making, the design, implementation, and evaluation of project activities in a manner that benefits them directly? Which individuals have taken the most important decisions which have contributed to shaping the project's future, and have these individuals been accountable to the intended beneficiaries — and how?

In asking these questions, the purpose is to draw PAF's attention to some of the issues which need to be examined if its claim to a participatory methodology is to be believed. The project has shown a considerable ability to listen to farmers, a preparedness to be flexible and to innovate. It has, however, been weak on its approach to issues of gender, especially the importance of targeting women in its work.

In attempting to draw up a balance sheet, attention must be paid to the reality of the project's short history. However, the history of PAF has so far not helped our understanding of the character of its participatory methodology. Its annual reports usually contain very little to be learned by other projects or individuals who want to benefit from its participatory methodology. The 1991/92 project year report, for example, was basically a narration of the activities undertaken, its achievements, and the problems that it encountered. For the year in question, the project lists the following factors which affected its work: differences in philosophies of intervening agencies; the frequency of exchange visits, with the project receiving four visits (from extension agents and farmers) between July and December, and eight visits a month between January and June; and the inadequate training of extension agents and farmers. Other factors are mentioned: problems of feeding confined animals, delays in the release of funds from Oxfam's head office, and the effect of national political campaigns in intensifying conflicts between villages. However, the report does not provide any detail about the specific ways in which these factors affected the project's work, and what plans it has made to cope with or solve them.

This lack of a detailed analytical view of the project's life for the year may well be due to the need to provide a short and simple summary of its activities, emphasising its quantifiable achievements. Such a document may well be the preferred approach for overseas funders more interested in seeing concrete results. Yet it could suggest that there has been no critical and evaluative look at the project's activities for the year. What an outside observer or interested party would like to know is precisely how the project coped with its problems.

Institutional learning

The problem with PAF, like many other NGO projects, is the lack of any documentation of experiences, written in a manner which facilitates the learning of lessons. There is little written information on where the project has come from, why it has arrived at its current state, and how this experience will inform its future. This problem is very much recognised by Oxfam. As Alice Iddi, formerly Deputy Representative of Oxfam in Ouagadougou, now the

Representative, points out: 'Oxfam-West Africa's performance evaluation is very weak. The programme as a whole has never been reviewed, and the few attempts to evaluate individual projects have been sporadic and narrow. Often, they have focused on project activities, not their context or relation to other important aspects of people's livelihoods. Many project evaluations are equivalent to financial auditing.'

Such a frank admission by an Oxfam staff member is certainly refreshing, and could serve as an example to other agencies. But the reasons advanced for this situation are disturbing: 'Like other NGOs, several factors have contributed to Oxfam's poor record in pro-gramme/project evaluation. These include inadequate knowledge and analysis of the social reality of the local environment and the wider systems that affect it; the focus on quick and concrete results to satisfy funding requirements; and practical difficulties of social-impact evaluation, such as the delayed or diffused effects of a soil and water conservation project on people and the environment,' Alice Iddi adds.

An explanation of this nature could cast doubt on the basis of the international acclaim that Oxfam, and its project PAF, have received. This would be unfortunate, since it may suggest that the acclaim is undeserved or that the achievements of the project have been exaggerated. Yet local farmers interviewed for this book did give credit to the project. Should we infer that Oxfam (and in this case PAF) is like a wanderer who accidentally discovers gold during an aimless journey and, in order to stake a claim of ownership, explains the find as the result of good navigational skills in what is undoubtedly difficult terrain? If Oxfam's activities in the West Africa region have not been informed by an adequate knowledge and analysis of the socio-economic reality of the region, then whatever progress Oxfam and PAF have made has perhaps been due more to luck than planning. If, however, these achievements have been based not on luck but on planning, then the question arises: what exactly has been the basis of the planning?

It must be pointed out, however, that while planning is important, it is illusory to present planning (even if it is strategic) as the magic wand for the solution of all rural problems. In certain circumstances, important results are achieved simply because of a readiness to try out new ideas which have not been critically analysed or planned for. Sometimes it is like trying out new seed varieties, simply in the hope that they could produce a good yield. Sometimes it works; sometimes it is a disaster. Yet in such situations, it is not so much the absence of knowledge which is the most important element, but the ability of a project to learn lessons quickly and adapt creatively to the new situation. In so doing, it may be necessary to go beyond the framework of an agreed and well-designed strategic plan. It is the ability of a project to learn the correct lessons which determines, to a large degree, the progress it will make.

Technological development

An absence of a 'critical knowledge' of the local environment, its potentialities and limitations, nevertheless poses certain dangers. One possible danger is the romanticisation of indigenous knowledge and the participatory methodology which, in the case of *diguettes*, is alleged to have aided its adaptation. Another danger is not knowing the real content of local skills and what it requires to improve upon them. Such over-estimation of the worth of indigenous knowledge means an exaggeration of the utility of local knowledge far above its real contribution. Such an exaggeration may well be useful for giving Oxfam and the project a unique identity. However, a failure to analyse critically the real content of indigenous knowledge means that there will be no systematic approach to developing local potential. So far, there is little evidence to suggest that the participatory methodology which guides PAF's research efforts has come up with anything effective, other than stone *diguettes* constructed with the water tube. Alice Iddi attributes the project's inability to stimulate farmers' own initiatives in technology development to a mutually reinforcing set of mentalities observed in villagers and project staff. The majority of villagers believe (or have seen the advantage of acting as if they believe) that those who have gone to school are wiser. As a result, they expect the educated outsiders to provide solutions to their problems.

Certainly, adapting and popularising the use of the water tube was innovative, but it depended more on the preparedness of technical staff to innovate than on an adaptation of indigenous methods of contour determination. There is nothing necessarily wrong with that, except when it creates an impression that technical innovation is an easy process involving a slight adaptation of local techniques. Idealising local knowledge can ignore the fact that it is often because indigenous knowledge has proved ineffective that intervention by a foreign agency is sometimes necessary (if often unsolicited). It is usually forgotten that learning scientific skills and methodology is as important as developing indigenous skills. There is a need, in the project's emphasis on participatory methodology, not to stop at using it to solve practical problems, but to guide farmers, where necessary, to experiment *in a systematic manner* with their own ideas about how to solve the problems which confront them. Local farmers must stop blaming the gods for most of their problems.

In saying so, it is not intended to glorify 'universal' science. Scientific methodology does not always produce useful results applicable to all circumstances. For example, scientific methodology produced Green Revolution technology, which led to the development of high-yielding and disease-resistant seed varieties. It did so by holding variables such as land, water, and income to be constant (though in various combinations), without taking into consideration the crucial issue of rural poverty. Its results have not proved beneficial to many farmers. The attempt to use Green Revolution technology to produce high-yielding sorghum and millet varieties adapted to the conditions of the Sahel and widely acceptable to farmers has so far been unsuccessful. Local farmers have had to rely on their own method of developing improved seed. This involves selecting the best seed from the annual harvest and storing it for use as seed grain in the next harvest. Through the Darwinian principle of natural selection, they have managed to produce seed varieties that are resistant to many of the variables which change in their environment. What this implies is the need for a combination of both traditional and scientific approaches.

Institutional structures for participation

The project documents' emphasis on its participatory character and methodology also raises the issue of what the project's peculiar features are. The project's organisational structure shows no particularly unique features. If there is nothing unique about the project's structure, then we must ask how an organisation can structure itself in such a way that the interests of local farmers could more easily define and shape the project's agenda. So far, there is little in the project's current structure which indicates that its claim to the involvement of the local community is any more legitimate than that of other agencies operating within the province.

PAF presents itself as particularly responsive to farmers' views. Some agencies such as 'Six S' claim to be a re-creation of traditional institutions, and to be following an agenda determined by local farmers. Such claims are sometimes based on the fact that they draw up annual programmes which reflect farmers' wishes and goals, collected at group meetings or through extension agents. Quite often, the fact that farmers' views are gathered is no guarantee that they will guide the project's activities. The views collected may be representative of local interests, but do local farmers actually have any managerial influence over the final shape and direction of the project?

PAF also emphasises 'awareness raising' and organisation as two of its strengths. Project staff in general, and the co-ordinator in particular, are known for their tireless efforts to make local farmers aware of improved land-use techniques and natural-resource management. The co-ordinator is renowned for his command and use of the Mossi language and traditional proverbs as a tool to impart new skills. Extension staff are also known for the amount of time they spend within their respective zones. Although this is usually indicative of dedication, its usefulness depends on the character of such visits. There is a thin line between teaching farmers 'new' techniques and encouraging them to develop upon what they know, using new ways that they learn. When is learning new skills a form of popular participation, and when is it a transfer of technology (teaching farmers new and presumed better skills)? What is the appropriate method for achieving the desired mix? Very often, extension agents see their role as educating farmers in how to use the water tube, construct *diguettes*, prepare compost pits, and grow fodder and trees on their farms. Farmer often wait patiently to receive the perceived wisdom.

Given the current structure of training and operation of the system of extension agents, this is unavoidable. However, encouraging farmers to build upon their knowledge may require a different approach to extension work. It may mean that the project structure and budget should incorporate the development of farmers themselves as extension agents. Perhaps current extension agents should be training farmers to be extension workers. Their role could be to share with farmers new and/or alternative approaches to solving problems and sharing the experiences of farmer/extension agents. Periodic workshops, using a combination of GRAAP and PRA techniques, could be a useful approach to achieving this. GRAAP techniques, developed in Burkina Faso by the Groupe de Recherche et Appui à l'Auto-Promotion Paysanne, consist of a series of visual aids and methods of communication developed from local imagery and communicative skills to impart new ideas. PRA (participatory rural appraisal) techniques emphasise the active involvement of peasants themselves in identifying and finding solutions to their own problems.

In their new role, farmer/extension workers would not need to be full-time paid employees. However, they would have to be financially compensated for the time that they travelled around the village sharing their skills with the rest of the community. This could prove cheaper than maintaining a large extension staff, as some NGOs do. For example, Village Groups could nominate one or two representatives to undergo training as farmer/extension agents. They would then return to their communities and perform their new role within a time frame agreed upon between the project and Village Group. After the agreed period, farmer/extension agents would stop receiving financial assistance and any active extension work that they did would be on a voluntary basis. A rotation of farmers for training as extension agents would be important, to ensure that as many people as possible were trained.

There are, however, certain dangers. The notion of temporary financial compensation for farmers for lost labour time could create enormous confusion within the rural community. The dire need for cash incomes could lead to competition and conflict over who was to undergo training. This could destroy the solidarity and cohesion of Village Groups, especially if selection of trainee farmer/extension agents was based on family ties, rather than on an individual's communication skills and support within the village. It could lead to a situation where Village Groups could quietly work out a system of rotation if they knew the programme would be prolonged. However, if development is a learning process, and if PAF has a record for trying out new ideas, then it has a responsibility to research new ways of ensuring effective participation of farmers in the project's work. This is necessary, since some concern has been expressed within PAF about the extent to which trained farmers have transferred their skills to their respective communities.

Village Groups and institution building

PAF emphasises organisation, preferably through whole

villages organised as a united community (to avoid rivalry between Village Groups) or through Village Groups where necessary. Working through Village Groups (which often involves only a section of the village) is meant to strengthen social organisation and facilitate the development of effective local institutions. Such institutions should be capable of mobilising their own resources to resolve their own problems. However, there is no clear indication of the progress that has been made by the project in this direction. Are Village Groups in pilot villages necessarily better organised because they work with the project? Will they continue to be well organised if the project stops working with them? The data do show that nearly 90 per cent of household heads sampled said they belonged to Village Groups; but this casts little light on the reason for the high reported levels of participation in group activity.

PAF aims to do everything it can to make the people self-reliant and independent of the project. Villagers do appreciate the help that PAF has given and continues to give. As a local farmer explained, 'We have mastered the *diguette* construction. We can continue to construct *diguettes* without PAF's supervision, provided they give us minimum equipment. We cannot buy our own equipment, because we have not got the means to do so. The work we are doing with PAF does not generate any revenue to allow us to do so.' Mahmoud Sawadogo of Ninigui argued that they were doing their best, and that it was not as if they enjoyed 'begging'. 'Our group charges a symbolic sum of 250 FCFA to anyone who wants to borrow, say, a cart for his private use. This money is used to repair the equipment. It is too small to buy new equipment with. That is why we rely on PAF for help. Nobody enjoys begging. We started preparing our fields with our own rudimentary tools: pick-axes. It is because we are motivated. It is our determination that made PAF give us better tools, but they are so far not enough. It is because we want to be more efficient that we are soliciting PAF's help.'

Yet there is no clear way of knowing whether Village Groups can manage on their own if no close scrutiny is given to the qualitative aspects of group development, and the capacity of Village Groups to handle their own problems. Who are the leaders of these groups, and do they represent the interests of the majority? Are Village Groups built around an individual (no matter his or her good intentions), in whose absence the groups will disintegrate? What is the strength of groups that look dynamic, and why are other groups not so active? PAF has tried to address some of these issues by occasionally organising inter-village competitions. However, although such competitions are useful in highlighting inter-village differences and potentials, the competitive atmosphere is probably not the best method for ascertaining the institutional strength of groups, and it can easily overshadow fundamental weaknesses.

So far, PAF has not developed an effective methodology to deal with institution building. There may never be any prescribed formulas, especially as particular local circumstances will determine the form that institutions take. For example, Mossi society is hierarchically structured, and the role of the chief is crucial. As the chief of the village of Noogo put it, 'It is the head which makes the snake. When the chief himself takes part in a job, the old ones know that it is serious and they turn up. And when the old men are mobilised around any activity, no one can stay away. That is how we mobilised our village for the *diguettes*.' Although the formal role of chiefs is changing, the mentality is slow to change, despite the impact of the Sankara years. Between 1983 and 1987, the role of chiefs in society was undermined by the prominent role given by government to mass organisations such as the Committees for the Defence of the Revolution (CDRs).

What the chief of Noogo forgot to mention was that the head of the snake carries the poison which kills. Centralised Mossi society does not have a tradition of changing leaders before their time is due. This situation can keep an inadequate leader in his post longer than is useful. More often than not, leaders get removed only when they become an obvious nuisance. Before that, any attempt to change a leader when there is no overwhelming support for the move can produce situations which lead to the collapse of the group. The membership can split in half, thereby paralysing the decision-making process.

Developing a methodology for building the institutional capacity of Village Groups is an area on which the project needs to conduct research, in conjunction with its partners. After all, the sustainability of any project activity depends greatly on local organisational capacity to continue with ongoing activities. Such a capacity can develop only if local groups are encouraged or supported to assume responsibility for managing activities run by the project. As a trial exercise, for example, an individual or a small team of Village Group representatives can be mandated to assume direct responsibility for running and supervising the project's activities for a period of time. After all, if institution building involves managerial skills such as decision making, the building of a consensus around clear objectives, implementing and monitoring the progress of work undertaken, and generating and allocating resources, it could be useful to find out what a village team would do with the equivalent of the project's quarterly budget for project activity. How will they go about defining their priorities and ensure that these are carried out? However, if a trial exercise is to yield any useful lessons, then it is important for project staff to give local farmers complete autonomy to decide the allocation of resources.

It is tempting to rule out such an exercise as a useless gimmick, considering the predominantly non-literate rural environment and the project's need to consult regularly with its main partners, the State agencies. Managing a project certainly requires reading and writing skills, but there is more to it than that. Although reading is important, it is writing of all the literacy skills which will be most needed if the exercise is to provide an opportunity for institutional learning — an aspect which the project clearly lacks. However, if 'modern' managers rely a lot on their secretaries, there is no reason why the team cannot be provided with an efficient secretary to document the experience.

Past, present and future

PAF's claim to cost-effectiveness

Another area which PAF identifies as its strong point is its ability to achieve much with limited resources. There is, however, little evidence to suggest that this claim is necessarily justified. It is probably based on a cursory comparison with the resources available to other projects in the province. The temptation to feel proud of having made some progress, despite not having the resources of other projects, can easily shift the argument away from the essence of cost-effectiveness. The issue, in determining a project's cost-effectiveness, is to find out whether the project's achievements are based on its ability to achieve results with diminishing resources. The most important element in such an exercise is the cost per head of each activity.

It is within this context that the validity of the project's claim to cost-effectiveness can be established. Table 6.1 covers the period 1982/83 to 1992/93. As the data stand, they provide no clear basis for determining the project's cost-effectiveness, more so since there has been no clear distinction between what goes into running the project (staff costs) and what benefits farmers more directly. Although there are major difficulties in trying to assess PAF's cost-effectiveness, for the purpose of this study an attempt has been made to compare expenditure used in running the project and that which we have categorised as probably benefiting local farmers directly. In making this distinction, we are conscious of the fact that extension work is a pillar of the project's activities and does benefit local farmers.

Table 6.1 suggests that for the first four years (1982-86) of its life as an operational project, PAF did not purchase any inputs to help farmers in the construction of *diguettes*. This is probably a reflection of the fact that its priority within that period was to train farmers in the technique of *diguette* construction, especially in the use of the water tube. However, in the 1986/87 financial year the project did budget for project equipment (£15,643) and £4,642 for its co-operation with State agricultural extension services. The

Table 6.1: PAF's budget – 1982/83 to 1992/93

Year	Total amount (£)	% of total allocated to farmers
1982/83	26,460	n.a.
1983/84	35,250	n.a.
1984/85	20,128	n.a.
1985/86	27,870	n.a.
1986/87	68,389	30
1987/88	93,277	37
1988/89	100,961	1
1989/90	122,520	n.a.
1990/91	140,686	16
1991/92	102,811	21
1992/93	127,126	27

n.a. = no specific figures available.
(Source: Oxfam, UK and Ireland)

expenditure that year which probably benefited farmers directly was 30 per cent of the total budget. The following year it increased to 37 per cent, but declined sharply to 1 per cent the subsequent year, mainly because the project froze its activities in anticipation of an external evaluation. For the 1990/91 financial year, the estimated allocation going to farmers directly was 16 per cent of the total budget. This increased to 21 per cent and 27 per cent in the 1991/92 and 1992/93 financial years respectively.

The figures show that on average, expenditure on running costs takes up about 70 per cent of PAF's budget. The high budgetary allocation for the cost of maintaining staff is understandable, since staff salaries, insurance and pension schemes, vehicles and office equipment do take a large share of any budget. This is perhaps why in many organisations job redundancies are usually the first managerial option when faced with decreasing financial resources. Furthermore, salaries in Burkina Faso are high compared with those in neighbouring countries. In the case of PAF, it is difficult to make a case for lower project costs unless there is a unit-cost-per-staff comparison with other agencies operating in the area.

Nevertheless, it is worth asking why, in the light of the pervasive poverty of the area, local farmers have not really benefited from the resources allocated to the project in a way that can be easily quantified. Could it be due mainly to the difficulty of quantifying the benefits to farmers of a soil and water conservation project? Perhaps yes, perhaps no. The figures suggest that while the project's overall budget has been increasing, this has not altered the basic project/farmers budget ratio. This could be explained by the wider range of activities that the project is engaging in, which therefore entails higher running costs. Yet as the preceding chapters showed, the basic problem facing farmers is their inability to acquire the inputs needed to translate the project's message into reality. As a farmer from Solgom-Nooré put it in relation to training given in the production of compost for fertilising farmland: 'Producing organic manure is one thing; how to get it to the site at a time you need it is another.'

Reaching the most disadvantaged

Oxfam's stated aim is to reach the disadvantaged in society — but in the case of PAF the rhetoric of this aspiration often seems stronger than the practice, if the budgetary allocation examined above is a useful yardstick. Yet the issue of helping farmers, who are meant to be the beneficiaries, to derive optimum advantage from Oxfam's support is not as simple as it appears. It may well be more expensive to reach the poor using a participatory approach, especially when calculated at cost-per-head. A quick way for farmers to benefit is for the project to purchase much-needed inputs and distribute them freely, or sell them at a heavily subsidised price to farmers. While this would quickly increase the volume of resources going directly to farmers and also eliminate the need for having many paid staff, its utility is extremely questionable. It would reinforce the aid-dependent mentality and destroy any potential that may exist locally to find solutions to problems whose scale and nature change frequently. The very poor households for whom such an exercise is designed could lose out eventually as some chose to sell their inputs to richer farmers for urgently needed cash. Such aid is certainly not sustainable, if even it were desirable.

There is also the fact that helping the most disadvantaged is not as easy as it sounds: we first have to define precisely who are they are. Within the context of Yatenga society, 'the poor' are invariably seen as the destitute, especially those who are physically disabled and lack family support. They certainly need special targeting. However, in targeting such a group, it may well be more appropriate to anticipate long-term support, since disabled people encounter more serious problems than able-bodied people.

The majority of rural households belong to the 70-80 per cent of the population who are neither destitute or rich. Within this category there are differences between and within households. There is also an on-going process of differentiation which requires a regular reappraisal of indicators of wealth. Patterns of inequality are influenced mainly by access to labour, land, production inputs, ownership of livestock, and cash incomes. All of these factors vary over time and between households. Therefore, even if Oxfam through PAF develops criteria for identifying poorer households and finds a precise way of assisting them to improve their livestock holdings, meet their labour requirements, and improve their sources of cash income, the process of social differentiation will continue — despite the contribution of the project.

The point being made here is that a detailed understanding of the social reality is indispensable if the project is to be effective. It is indispensable because it could help project staff to understand the underlying factors which direct the process of social differentiation, and the particular importance of each factor. This can enable them to identify measures more appropriate for reaching the disadvantaged. Despite its indispensability, it is important to bear in mind that in practical terms it is not always easy to design activities which benefit only the poorer sections of rural society. In any case, it could also be argued that doing so may well lead to the poor being exploited by the rich.

Furthermore, considering the environmental threat facing the region, it is worth asking if halting the tide of environmental degradation should not take precedence over concern to reduce rural inequality. This is not to suggest that bridging the gap between the rich and the poor is unimportant. On the contrary, it has to be borne constantly in mind. However, the project should not feel guilt-ridden, because in many cases the reinforcement of inequality is unavoidable and can even yield some positive benefits, such as offering employment opportunities to the poorest. For example, as Iddrissa Sawadogo recalls, 'I was among the pioneers who learned how to construct *diguettes* from PAF. Seydou asked me to prepare his fields, and he brought lots and lots of truck-loads of stone. Three young men from my village and I worked on his farm and he paid us 50,000 FCFA. Since then, no matter how bad the season, Seydou always manages to harvest something.' Seydou has undoubtedly benefited from the technique more than the young men he hired. His harvest will improve significantly, if other factors are favourable. But there should also be a positive impact on the Yatenga environment.

All this goes to show the complexity of rural society. Yet unravelling this complexity is useful in any attempt to shape the project's future.

Looking into the future: PAF and the State

As PAF charts its future role in Yatenga society, it is looking to a future that presents both opportunities and dangers. In its 1992/93 to 1996/97 programme of transition, the project outlines its plans mainly within the framework of the National Programme of Village Land Management (PNGTV), initiated in 1986 following the promulgation of the agrarian and land reform in 1984. As its major areas of emphasis for the next four years, PAF has identified land-use management, intensification of the struggle against soil erosion, and the promotion of agro-forestry; an intensification of animal production; and an improvement of the economic environment within pilot villages. It hopes to achieve its objectives through an increased supply of equipment and materials to farmers, the continued use of the revolving stock scheme to help poorer households, and an intensification of its participatory research activities. It hopes to carry this out in closer collaboration with State structures operating within the context of PNGTV, now referred to as the National Programme of Land Management (PNGT).

The National Programme of Land Management (PNGT)

PNGT was first mooted under the regime of President Sankara. Intended to be an integral part of the proposed agrarian and land-reform programme, PNGT was designed as a participatory multi-sectoral and decentralised development effort, assuming the village community to be the basic and most important unit of decision making about the usage of natural resources. The programme was built on the premise that natural resources were common property and should be developed collectively for common use. As a result, all lands were vested in the State and could not be appropriated without the State's approval. This measure, as far as the rural areas are concerned, was meant to control the development of uncultivated land, since in practical terms it is impossible to nationalise peasants' farmlands.

The agrarian and land-reform programme and its integral

programme, the PNGT, were conceived with the attainment of food self-sufficiency and improved living conditions in mind. This was to be done mainly through a 'rational' exploitation of village land resources. A rational land-use system was to involve the protection and conservation of the environment and the organisation and utilisation of village resources — soil, water, and vegetation. The utilisation of village land resources was to be based on a village development plan compiled by the local community in conjunction with State agencies and other actors within rural areas. A key element in a village development plan was to be the security of tenure of village landholdings and the control by the community (as a whole) of its resources. It was expected that since the village community would be playing the leading role in the development process, they would be in a better position to ensure that the programme met their needs. Although the circumstances surrounding the promulgation of the agrarian and land-reform law changed after the dissolution of the National Council of the Revolution (CNR) in October 1987, the government of President Blaise Compaoré declared its commitment to the programme. Consequently, it remains the basic framework for rural development and has been supported by international financial agencies such as the World Bank and the IMF.

The future of PNGT

As we have seen, PNGT was conceived during the sweeping changes which followed the advent of Captain Sankara and the CNR to power. It was hoped that the climate of popular mobilisation would provide a suitable atmosphere for an integrated approach to development. Although various aspects of the programme involved different ministries, they were expected to co-ordinate their activities in the field. However, a recent review by the World Bank has argued that, in reality, there has been little effective co-ordination between the various government departments at the rural level. Furthermore, since the programme was essentially a top-down approach to development, village communities have not responded in a manner which suggests that they see themselves as the pillars of rural development. Moreover, the programme was based on the false assumption that it is always possible for an entire village to exploit its natural resources with regard to the broader interests of the whole community. Yet real-life experience suggests that richer members of the village community, with large livestock holdings, would be more interested in reserving wide tracts of land for pasture. Those without significant livestock, but lacking in farmland, would be more interested in gaining access to cultivable land. In the absence of political leaders committed to protecting vulnerable people, collective village management of resources could benefit mainly the richer households. The concern of the Bank and the IMF is, however, centred on the agrarian and land-reform law which sought to give collective access to village land resources. In the light of the changing political climate, especially the establishment of multi-party politics, the concerns expressed by the Bank and the IMF, and their insistence on changes to the agrarian reform law, the indications are that PNGT could become another rural development programme with good public relations, but minimal impact on the life of rural communities.

The World Bank insists that unless there is a change in the agrarian and land-reform law which gives collective access to village land resources, the programme will fail. In the view of the Bank, it will fail because private individuals, with no security of tenure, will not be encouraged to invest in activities which can improve the land. Also, an agreement between the government of Burkina Faso and the IMF for a structural adjustment loan, signed in 1991, made proposals for major reforms in the agricultural sector which could have serious implications for any NGO operating at the village level. The draft discussion document for negotiations with the IMF proposed a major disengagement of the State from directing the process of agricultural development.

The reforms plan to tax rare land resources (whatever this means) in order to encourage farmers and herders to improve their management of natural resources, and make the necessary investment needed to increase productivity. Accompanying this policy will be a gradual and progressive disengagement of the State from service provision to rural farmers. Provision of services is expected to be undertaken by Village Groups and producers organised on a national scale. It is expected that the disengagement of the State from service provision could make Village Groups more dynamic and enable them to take overall responsibility for the distribution of agricultural inputs, to develop drug stores for animal products, and to assume responsibility for animal vaccinations.

However, the very reforms which the Fund claims could dynamise Village Groups could in reality kill off grassroots initiatives by poorer sectors of rural communities to improve their productive capacity. Introducing a tax on rare land resources will probably deprive poorer sections of the community of access to what has been known traditionally as common property. Furthermore, Village Groups tend to have a limited capital base. Those that were formed during the era of popular mobilisation came into being with the understanding that financial contributions were not to be the main feature which identified them as socio-economic groups, but rather their cohesiveness and ability to represent and promote the interests of the village. Also, Village Groups formed before 1983, as Lédéa Ouédraogo of the 'Six S' movement has argued, tended to be weak institutions, since obtaining inputs had been the *raison d'être* for their existence. To entrust the provision of inputs to such groups will in real terms hand over control of such inputs only to those who can afford to purchase them — the richer sectors of peasant society. A disengagement of the State from service provision (such as extension support to farmers and pastoralists, and construction of basic infrastructure), coupled with the freeing of price controls on agricultural inputs, could prevent under-resourced farmers from obtaining much-needed equipment on favourable credit terms. Moreover, what could put the final nail in the coffin of rural farmers is the proposed plan for economic liberalisation, where the importation of all agricultural inputs and products will be freed from State control. Against this background, it is debatable if the intended goals of the reforms (a revival of animal traction, improved inputs, and a supply of agro-industrial products to pastoralists on the basis of opportunity costs) will actually lead to a modernisation of the agricultural sector. On the contrary, they are likely to modernise only the already modern 'model farmers', and bypass the vast majority of subsistence producers in the peasant sector.

There are other aspects of the proposed reforms which could prove damaging. Abandoning the State's monopoly over the importation of agricultural inputs and products (except for rice, sugar, and oil) promises to open up local agricultural products to competition from cheap foreign imports. In Yatenga, where vegetable market gardening is an important source of cash income, free-market economics is probably not the best way to protect the interests of rural farmers. Unless there is some protection and State support for Yatenga's vegetable industry, the hope that vegetable growing has generated an alternative source of cash income could vanish. It is also doubtful if the scheme to introduce a system of taxation on rare land resources can be implemented. Considering the fact that head taxes on peasants were abolished under the government of President Sankara, and that the present government of Prime Minister Ouédraogo and his party, ODP/MT, consider themselves legitimate heirs to processes which began in 1983, an attempt to reintroduce any form of rural taxation could severely undermine the political legitimacy of the present government. Furthermore, there are the practical difficulties of collecting rural taxes.

There is yet another aspect which could undermine the basis of the PNGT programme — and thereby the ability of PAF to achieve its objectives. When the planned changes in land-use patterns were initially mooted, there was a conviction that popular momentum generated under Sankara would facilitate their implementation. The most important obstacles were identified as changing rural attitudes and raising peasants' consciousness and mastery of 'new' farming

Past, present and future

skills. As a result, the operational strategy was identified as consciousness raising, dissemination of information, and the training and education of village communities in what was expected of them under PNGT.

In an atmosphere of generalised mobilisation, the task of raising rural awareness and promoting changes in existing patterns of land utilisation is comparatively easy. Popular support for a government is an important requirement for mass mobilisation and provides a receptive climate for new ideas. Awakening rural consciousness of the need for a 'rational management of land resources' is important if rural communities are to be directly responsible for implementing the programme. A harmonisation of the interests of different sections of the rural community with those of political and administrative structures and representatives of State technical services would give the programme a broad democratic foundation.

However, PAF's annual report has already indicated that inter-village conflicts, emanating from political rivalry, were an important factor which affected its activities during the 1991/92 farming year. Such conflicts often have a long history. Inter-village conflicts are rooted in the complex land-use system and structures of social organisation. (For example, in the village of Recko, there was a conflict brewing in 1991 over animal confinement. The village chief and his family were opposed to it, while the Village Group was in favour of it. Yet the conflict was not really based on disagreements over the principle of confining animals. It had its roots in the period of Sankara (1983-1987), when a disagreement erupted over the election of the village CDR (Committee for the Defence of the Revolution). The chief and his family felt that the delegate should have come from the royal line, but in the prevailing revolutionary climate the family's wishes were overruled. Now, apparently, they want to get their revenge on the Village Group, and are using the issue of animal confinement as a pretext.)

In an atmosphere of electoral politics where most of the parties are seen by their leaders as vehicles for negotiating with the State for resources, the danger of the programme's becoming an electoral tool at the village level is a real one. Electoral politics, with its tendency to promise what cannot be fulfilled, and to fulfil only that which secures majority votes, can produce Village Development Plans which are essentially meant to strengthen the political power-base of political aspirants and their allies within the rural community. It is within the above context that PAF's contribution to development activity within the framework of PNGT will take place. Yet its future will also be shaped by factors beyond the scope of the programme of village land-management.

Conclusion

It is still early days to make any confident predictions about the future of PAF. Its technical achievements over the past decade are undeniable: soil-moisture levels in its pilot villages have demonstrably improved; cereal crop yields have increased; a wider diversity of non-cereal crops is being grown; natural vegetation is spontaneously regenerating; farmers observe that the agricultural cycle on fields treated with diguettes is slowly returning to the way things were before the drought and famine of the 1970s. The social and economic benefits of PAF, however, are less easily demonstrated, and perhaps more time must elapse before we can make a final judgement. At all events, whatever modest successes PAF is able to claim are now at risk from political pressures that originate beyond Yatenga.

The prospects for the future of the National Programme of Land Management (PNGT), summarised in the previous section, do not bode well for PAF's planned activities, in the light of the projected links between the two. Despite the integrated approach that PAF intends to adopt through its collaboration with State agencies, interdepartmental co-ordination may mean, for a long time to come, more frequent meetings between bureaucrats, rather than a harmonisation of various departmental support structures at village level.

Past, present and future

Furthermore, State support in terms of the provision of basic socio-economic infrastructure will not be so readily available as was originally envisaged when PNGT was first mooted. This means that PAF will be confronted with the same problems that it has encountered over the years: lack of easily accessible drinking water, poor health and educational infrastructure, and the absence of alternative sources of cash incomes for rural farmers. The project's objective of targeting the less advantaged in segments of the local community will most probably be undermined by the attempt by international agencies to privatise land resources. It will equally be undermined by the village groups' assumption of full responsibility for distributing inputs, because that will simply provide an opportunity for wealthier farmers to capture the market for farm implements.

Despite what is probably an unfavourable institutional climate for ensuring that the poor are the main beneficiaries

Planting for an uncertain future: Oxfam's Daouda Ouédraogo teaches children in Goumba village to plant acacia seedlings along a diguette. But will marginalised communities develop their own institutions, capable of negotiating for their share of national resources?

of the project's efforts, the future of PAF will depend even more on whether and how it restructures itself to achieve greater involvement of local farmers, more effective planning ability, and better documentation of its experiences. Its future will depend, too, on the evolution of national politics during the era of political pluralism, especially if the latter provides an opportunity for vulnerable sections of rural communities to make effective demands on the State for increased allocation of national resources to rural areas. Political prioritisation of marginalised communities could in turn depend on how village-level associations develop to become effective socio-political development institutions, capable of co-ordinating their efforts at the local level and organising from a united front to put pressure on political leaders.

References and further reading

Chambers, R., A. Pacey, A., L.A. Thrupp (eds.), 1989, *Farmer First: Farmer Innovation and Agricultural Research*, London: Intermediate Technology

Conroy, C. and M. Litvinoff (eds.), 1988, *The Greening of Aid: Sustainable Livelihooods in Practice*, London: Earthscan

Cross, N. and R. Barker (eds.), 1992, *At the Desert's Edge: Oral Histories from the Sahel*, London: Panos/SOS Sahel

Farrington, J. and A. Martin, 1998, *Farmer Participation in Agricultural Research: A Review of Concepts and Practices*, Agricultural Administration Occasional Paper No. 9, London: Overseas Development Institute

Gubbels, Peter, 1992, 'Farmer-First Research: Populist Pipedream or Practical Paradigm? Prospects for Indigenous Agricultural Development in West Africa', unpublished MA dissertation, School of Development Studies, University of East Anglia

Harrison, Paul, 1987, *The Greening of Africa*, London: IIED/Earthscan, Paladin

Iddi-Gubbels, Alice, 1992, 'The Role of Northern Non-Governmental Organisations in Sub-Sahara Africa in the 1990s: The Case of Oxfam UK in West Africa', unpublished MA dissertation, Centre for Development Studies, University College of Swansea

Lamachère, J.M. and G. Serpentier, 1988, 'Valorisation

agricole des eaux de ruissellement en zone soudano-sahélienne. Burkina Faso — Province du Yatenga — Région de Bidi', Ouagadougou: ORSTOM

Ministère de la Planification et du Développement Populaire (MPDP), 1986, 'Premier plan quinquenal de développement populaire de la province du Yatenga', Ouagadougou: MPDP

Pacey, A. and A. Cullis, 1986, *Rainwater Harvesting: The Collection of Rainfall and Run-off in Rural Areas*, London: Intermediate Technology

Porgo, Z.P., 1992, 'Rapport de Stage: Technologie de la conservation des produits maraichers au Yatenga. Cas de Baas-neere'

Reij, Chris, 1987, *The Agroforestry Project in Burkina Faso: An Analysis of Popular Participation in Soil and Water Conservation*, London: IIED

Rochette, M.R., 1989, *Le Sahel en lutte contre la desertification: leçons d'experiences*, CILLS, GTZ

Wright, P. and E.C. Bonkougou, 1986, 'Soil and water conservation as a starting point for rural forestry: the Oxfam project in Ouahigouya, Burkina Faso', *Rural Africana*, 23-24: 79-85

Index

afforestation 35, 58-60, 137
AFVP (Association of French Volunteers for Progress) 28, 29, 30, 57, 70
Aly (of Gourcy) 24
animal confinement 93-5, 102, 105, 159
Armadou, Valian 25
Arnold, Corinna 30

bagtanga soil type 8-9
baobab tree (*toega* or *Adansonia digitata* 4, 5, 76, 124
bissiga soil type 8
Blade, Arlene 34-5, 36
Bogoya 31, 49
Boureima, Marcel 25
BSONG 25, 29, 66
bunds xiv, 27, 35, 63
Burkina Faso: administrative centres xxiv; agriculture xiii; external revenue xii-xiii; infrastructure xiv; population xii, 109; population density 109; secondary sector xiii-xiv

cash crops 4, 118
cash income 110-13
CATHWELL (Catholic Relief Services) 27
cereal deficits 107-8, 109, 115-16
CIEPAC 39
Compaoré, Halidou 47
Compaoré, Iddrissa 23
compost production 58, 70, 91, 92-3, 102, 104, 105
contour-line determination 38, 39-43, 84-6, 134, 137
cost-effectiveness xviii 149-51
Côte d'Ivoire 22-3, 95, 120, 132
cotton 9, 76

cowpea 8
credit 31, 70, 100-1, 107
crop diversification 75-6
crop yields 75, 117
CRPA (Regional Centre for Agro-Pastoral Development) 12, 56, 90, 104, 109

dams 30, 120
deforestation 2, 4, 39
diguettes: crop diversification 75-6; earth 26-7, 87, 103; evidence from farmers 72-5; maintenance 82-3; pitta planted along 60, 82, 83, 87, 103; plant growth along 76; run-off water 116-17; stone xiv, xv, 35-6, 72-84, 84-91, 102, 124-6
domestic water supplies 18-20, 117, 130, 131
droughts 2, 27, 110

educational provision 20-2
emigration 22-3, 120, 132
environmental degradation 2, 4, 152
erosion 8, 26
eucalyptus 59, 77
expatriate staff 49-52, 67, 69, 143-4
extension agents 144-5

famines 2, 109-10, 132
farming system 9-10, 60, 74-5
Fatimata (of Gourcy) 127
Fatimata (of Ranawa) 5
FDR (Rural Development Fund) 26
fertilisers 10, 60
Five-Year Development Plan 1991-95 2
fodder 93, 105
food 5, 110-11, 123-5

foreign intervention 49-50
fuelwood 4, 5, 122-3

GERES (European Society for the Restoration of Soil) 26-7, 37, 71
German Agro-Action Scheme 29
German Programme for Sahel 29
German Voluntary Service Overseas 29
Ghana 22, 95
gold panning 110, 111-12
Goumba village 8, 11, 54, 76, 124, 161
Gourcy 14, 16, 18, 21, 24
GRAAP techniques 144
grass barriers 38, 60, 82, 83, 87, 103
groundnuts 8, 13, 75, 130
Groupements Naam 32-3, 62, 67
Guiro, Kassoum (of Boulounga) 125, 126
Guiro, Mahama (of Goumba) 54, 72-3, 78, 124

Hamidou (of Ranawa) 5
handicrafts 14, 112, 113-14, 128, 130, 131
Hannequin, Bernadette 130
health services 15-18
Hereford, Bill 34, 101

Iddi, Alice 139-40, 141
Iddrissa (from Gounga) 124-5
Iddrissa (from Gourcy) 14
indigenous knowledge 141-3
indigenous staff 67, 69, 143-4
infrastructure xiv, 15, 161
inputs 106, 158
institutional learning 139-41
International Monetary Fund 17, 156-7

Kaboré, Marcellin 31
Kaboré (of Gourcy) 18
Kagoné, Hamidou (of Ninigui) 93
Kalsaka village 12, 112
Kanawa village 78
Kao village 49
kapok (*voaga* or *Combretum micrantum*) 4
Kombi Naam 32, 33
Koriga village 49
Koumbri district 5
literacy 20-1

Longa village 11, 12, 27, 54, 74, 76, 83
LUCODEB (Struggle Against Desertification in Burkina Faso) 57

Mady, Kinda (of Simbissigui) 73
Magrougou 8, 76, 91-2
Mahama (of Goumba) 134-5
maize 9
Mali 22
manure 91-2, 93, 105
market gardening 14, 114, 118-20, 128, 131, 158
Martin, Pierre 39
migration 22-4, 120-1, 132
millet 8, 9, 12, 14, 75, 110, 112
mimosa (*zaaga* or *Acacia albida*) 4
Ministry of Environment and Tourism (SPET) 56
Ministry of Planning and Co-operation 66-7
Ministry of Primary Education and Mass Literacy 20
Ministry of Rural Development 31
Mogom village 12

neem 59
néré (*roanga* or *Parkia biglobosa*) 4, 5, 76
Ninigui village 5, 8, 47, 67-8, 93, 146
NGOs 24, 25-52; co-operation between 70-1; competition between 66-8, 69; differing priorities 68-9; institutional learning 139-41; learning from past 26-7; multi-agency intervention 69-71; recent involvement 27-34
Noogo village 8, 27, 54, 78, 110, 122, 127

okra 9, 75, 76, 124
oleaginous crops 75, 76
ORD (Organisation of Rural Society) 43, 57
Ouahigouya 2, 15, 16, 21, 46, 110
Ouédraogo, A5seta 125-6
Ouédraogo, Daouda 117, 161
Ouédraogo, Hamidou (of Recko) 73, 92
Ouédraogo, Haouna (of Goumba) 24

Index

Ouédraogo, Harouna (of Ranawa) 2, 22-3, 120
Ouédraogo, Issoufou 27
Ouédraogo, Kassoum 73
Ouédraogo, Lédéa 32, 33, 34, 67, 81-2, 157
Ouédraogo, Mahama (of Longa) 74
Ouédraogo, Mahama (of Magrougou) 91-2
Ouédraogo, Mathieu (of PAF) 36, 43-4, 46, 49, 54-5, 66, 100, 107
Ouédraogo, Ousmane (of Noogo) 27
Ouédraogo, Ousmane (of Noogo) 78
Ouédraogo, Ousseni (of Ranawa) 42, 85-8
Ouédraogo, Sarata (of Longa) 76
Ouédraogo, Souleymane (of Kanawa) 78
Ouédraogo, Yacouba (of Sologom-Nooré) 86-7
Ouédraogo, Youssouf 54

PAE (Agro-Ecology Project) 28, 29-30, 67
PAF *see* Projet Agro-Forestier
participatory methodology 136-9, 143-9
pilot villages 64, 65, 75, 76, 90, 98, 104
pitta (*Andropogon gayanus*) 26, 38, 60; 76, 82, 83, 87, 103
PNGT (National Programme of Land Management) 154-60
Porgo, Lizeta 100-1
Projet Agro-Forestier (PAF) xiv-xv; adaptation of plans 37-8; agencies with which co-operates 56-7; animal confinement 93-5, 102, 105; areas of activity 57-64; awareness raising 143-4; choice of Yatenga province 38-9; collaboration with CRPA 104; compost *see* compost production; cost-effectiveness claim 149-51; creative adaptability 44; *diguettes* 84-91, 102; early years 34-46; evaluation xv-xix; expatriate staff 49-52, 67, 69, 143-4; extension agents 144-5; extension of work 95-6; flexibility 47-9; group activity promotion 54-5; impact on women *see* women; implementation of project 43-4; input provision 62-4; institutional learning 139-41; institutional structures for participation 143-9; intervention zones 3, 64-6; involvement of local farmers 37; limits to role 117-21; loss of interest 49-50; market gardening *see* market gardening; migration and 120-1; modest intervention 46-7, 53; objectives 53-4; origins 34-7; participatory methodology 136, 137-9; research and development 64; results of first operational phase 45-6; return on investment xviii; revolving grain stock scheme 61, 62, 98-101, 107; scope today 53-71; socio-economic impact 114-17; soil improvement 58, 91-3, 104; State and 154-60; strength of foundations 101-9; sustainability 134-5; targeting 132-4, 151-4; training 60-1, 95-6, 102, 104, 105-6; Village Groups and 96-8, 145-9
'Projet Micro-Parcelles' 35
PVNY (North Yatenga Food Growing Project) 28, 29, 31-2, 56

rainfall 11-14
Rana village 130
Ranawa village 2, 5, 8, 12, 22, 85, 120, 127
rassemponiga soil type 9
reafforestation 47-8
Recko village 73, 83, 92, 159
revolving grain stock scheme 61, 62, 98-101, 107
rice 9
Rimaibé 2
run-off water 26, 116-17, 37-8

Sanaa, Hawa (of Goumba) 76
Sankara, Abdoulaye 87
Sankara, President Thomas 15, 147, 155, 158

Sawadogo, Boureima 1, 4
Sawadogo, Fati 123
Sawadogo, Iddrissa 113, 153
Sawadogo, Mahmoud (of Ninigui) 146
SDID (Desjardins international Development Agency) 28, 29, 30-1, 70
sesame 9, 75
sheanut tree (*tanga* or *Butyrospermum parkii*) 4, 5, 14, 76, 130
Simbissigui village 73
'Six S' Movement 29, 32-4, 43-4, 67, 70, 98, 117, 143, 157
soil degradation 5-6
soil erosion 26
soil improvement 58, 91-3, 104
soil productivity 58
soil types 7-9, 82
Sologom-Nooré village 8, 86
Song-Song Taaba 62
sorghum 8, 9, 14, 36, 74, 75
State intervention 154-60
stones 38, 63-4, 79, 81-2, 102-3, 121-2, 124-6, 133-4
survival strategies 109-14, 118-20
sustainability 134-5

tamarind (*pusga* or *Tamarindus indica*) 4
Tambao 15
Tanghin Baongo village 8
Tao, Bintou (of Magrougou) 76
Tapsoba, Omar 30
targeting 132-4, 151-4
technological development 141-3
terracing 38
test villages 65, 75, 90, 104
Tiou village 12
Titao village 12, 16, 21, 30
tools 14, 63, 79
Tougri 127
training 60-1, 84-91, 95-6, 102, 104, 105-6, 139-41
transportation 63-4, 79, 81, 92, 105, 121-2, 126, 133-4
trees 5, 35, 47-8, 58, 59, 70, 127
Tugiya village 49
vegetables 75, 76, 119 (*see also* market gardening)
vegetation 4-5, 7
Village Development Plans 159
Village Groups 33, 50, 67; extension agents 145; GERES and 26; leaders 147; PAF promotion of 54-5, 96-8, 145-9; participants 138; pooled resources 106; provision of services by 157; revolving grain stock scheme 61, 62, 98-101, 107; rivalries 30, 134, 159-60; seedling trees 58, 59, 70, 127; solidarity 106

water resources 11-14
water sources 18-20, 117, 126, 130, 131
water tubes 39-43, 84-6, 134, 137
wells 19, 126, 131
women: *diguettes* construction 124-6; education 21; farming own fields 128-9; food availability 123-5; fuel wood 122-3; gold panning sites 111-12; grinding mills 31, 130-1; handicrafts 112, 113-14, 128, 130, 131; impact of PAF 121-32; income generation 127-9; increasing workload 121; independent income 129-30; market gardening 128, 131; midwives 17; need to target 130-2; participants 138; petty trading 31; transportation of stones 121-2, 126; water access and collection 19, 130, 131
wood 4, 5, 122-3
woodstoves (improved) 5, 122, 123
World Bank 26, 156
Wright, Peter 36, 39, 44

Yatenga province 1; domestic water supplies 18-20; educational provision 20-2; environment 2, 4; farming system 9-10; health services 15-18; migration from 22-4; PAF zones of intervention 3; population 2; relief 5, 6; roads 15; seasons 12-14; soil degradation 5-6; soil types 7-9; vegetation 4-5, 7; water resources 11-12
zay holes 46, 58, 82, 91, 103
zecca soil type 7-8, 82
Zida, Gabriel 29
zippela 72